T0261647

The

SOCIAL

AMOEBAE

The
SOCIAL
AMOEBAE

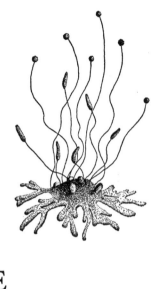

The Biology of Cellular Slime Molds

John Tyler Bonner

PRINCETON UNIVERSITY PRESS
Princeton & Oxford

Copyright © 2009 by Princeton University Press

Published by Princeton University Press, 41 William Street, Princeton, New Jersey 08540

In the United Kingdom: Princeton University Press, 6 Oxford Street, Woodstock, Oxfordshire OX20 1TW

All Rights Reserved

LIBRARY OF CONGRESS CATALOGING-IN-PUBLICATION DATA
Bonner, John Tyler.
The social amoebae : the biology of cellular slime molds / John Tyler Bonner.
    p. cm.
Includes bibliographical references and index.
ISBN 978-0-691-13939-5 (hardcover : alk. paper)
    1. Acrasiomycetes.   I. Title.
QK635.A1B665   2009
579.5′2—dc22   2008020569

British Library Cataloging-in-Publication Data is available

This book has been composed in Adobe Garamond

Printed on acid-free paper. ∞

press.princeton.edu

Printed in the United States of America

10 9 8 7 6 5 4 3 2 1

CONTENTS

Preface

I have lived with my beloved slime molds for a long time, and now suddenly I find myself quite overcome by the vast amount of new facts that have accumulated to account for every stage, every step (however small) of their life cycle. In the late 1940s and early 1950s (1945 to 1951) an average 3.4 papers on cellular slime molds were published a year; now, over the past seven years, there is an average of 224 papers a year! We are in danger of drowning in facts.

This is what stimulated me to write this book. It is an exercise to clarify my own thoughts. As I wandered into thinking about it, I decided to write something that is aimed at the curious layperson if I could explain the complete picture to him or her then I would understand it myself. I plan to shed all unnecessary jargon and concentrate on what I think are the main issues, all the while trying to give a rounded and complete picture of the biology of cellular slime molds. At the same time I will ask where slime molds fit into all of biology and even how they might illuminate that vast subject.

This will be done by concentrating on all of their biology, including their molecular biology. I plan to give equal time to their evolution, their ecology, their behavior, both as single amoeba and as cell masses, and their development. Some of these subjects have made great advances using molecular techniques; others, by their very nature, have been virtually untouched by molecular biology.

It must be said right from the beginning that this is a very personal book: it is my view of slime molds, and my view of biology. Others would no doubt write a very different book, but I hope that my individual slant does not interfere with the story I want to tell. My prejudices will affect the emphasis, and often the choice of topics, but my wish is not to obscure the significance of cellular slime molds and all the lessons they have for us.

I will look at everything through the lens of a full-time biologist. That will include the contributions that molecular biology has made that illuminate basic living processes. There is an enormous difference between what was known when I published the second edition of my *The Cellular Slime Molds* in 1967 and what we know now: it is quite overwhelming. One need only look at Richard Kessin's splendid book *Dictyostelium,* published in 2001, to again see in a dramatic way how rapidly the subject has advanced since 1967. And there have been many more new discoveries since his book was published.

The advances have been largely in the new molecular facts, and this is the reason there has been such a rapid turnover. It is my hope that, by looking at everything from a more generalized biological point of view, the lessons we have learned will last for some time.

One other difficulty in writing such a book is that there are scores of first-rate contributions that will not receive the credit, or even the mention, they deserve. This book is not an encyclopedia, a textbook, or a monograph; it covers only a fraction of all we know about slime molds. Rather, it is an essay on the big lessons we have learned, with a few unconventional and new insights. The details have been shed, fascinating though many of them are, and only the main plot is bared. Because of this I apologize to the many workers in the field whose fine work I have not mentioned, and some of them are—or at least were—good friends!

None of the facts in this book are new: many are common knowledge and some may lie unnoticed under a rock. If there is any novelty it is in the way they are put together.

A number of kind friends were good enough to read earlier versions of this book and their corrections of errors and advice for improvements were of incalculable help to me. In particular I would like to thank Leo Buss, Edward Cox, Vidyanand Nanjundiah, and Pauline Schaap. I would also like to thank Mary Jane West-Eberhard for looking at one section of the book. Throughout the preparation of this book Slawa Lamont has sustained me with her encouragement and her support. Finally, I have been lucky enough to get the tremendous, skilled help from a number of individuals at the Princeton University Press; my special thanks to Alice Calaprice, Alison Kalett, Dimitri Karetnikov, and Deborah Tegarden.

The

Social

Amoebae

1 INTRODUCTION

Evolution, cell biology, biochemistry, and developmental biology have made extraordinary progress in the last hundred years—much of it since I was weaned on schoolboy biology in the 1930s. Most striking of all is the sudden eruption of molecular biology starting in the 1950s. I will make a reckless generalization that each one of these surges was due to a collision with genetics. Perhaps it would be more accurate to say that they fused rather than just collided, because in each case an extraordinary fruitful symbiosis was the result. First, at the beginning of this century, genetics fused with nineteenth-century cytology, which gave us an understanding of how the genetic material was handled in the chromosomes in mitosis, and particularly in meiosis. Next, genetics fused with Darwinian evolution to give rise to population genetics, a signal advance at the time. Then, with the revolution started by Watson and Crick on the molecular structure of the gene, it was possible, through molecular biology to (1) have a second fusion of genetics with cell biology, making it possible to dissect out the biochemical or molecular events within a cell; (2) to devise a new way of attacking phylogenetic problems in the study of evolution using molecular-genetic techniques; and finally (3), these new approaches made it possible to dissect out the sequence of molecular steps in development.

I am not done with my generalizations! During all the events I have described, there has been a strong tendency to concentrate on "model" organisms. In the past century there was a tremendous emphasis on *E. coli,* the fruit fly *Drosophila,* and the nematode worm *Caenorhabditis,* but beginning back into the nineteenth century there have been many others that have played a similar role. To mention a few, there was Mendel with his garden peas, followed by other organisms such as maize, amphibian, chick and sea urchin embryos, yeast, myxobacteria, zebra fish, the small plant *Aribidopsis,* and the cellular slime molds. One could add a few more and the list would still be incomplete: for instance, ciliate protozoa (such as *Paramecium*), *Hydra* and other hydroids, sponges, *Volvox* and other algae, myxomycetes or true slime molds, *Phycomyces* and other fungi, mice and other mammals. The degree to which these various examples have been directly affected by genetics and molecular biology varies, but even in those cases where the influence has been small (due to the lack of attention) this is beginning to change. In fact, one can say that it is inconceivable to study the biology today on any organism without genetics and molecular biology. For completeness it should be added that there is now renewed interest in another collision: the realization that evolution and developmental biology are inseparable, something that was already recognized by Darwin.

My reason for using cellular slime molds as an example is that they (along with myself) went through the same evolution from a pregenetic period to one deeply involved in molecular genetics. I have been there to watch every step.

When, as an undergraduate, I began experiments on these slime molds in 1940, only one other person, Kenneth Raper, was working on them at that time. In fact, he discovered the "model" species *Dictyostelium discoideum*, which is the species used in the majority of the experimental work today. His early experiments were in the classic mold of the embryology of that time and are still recognized today as being at the root of all subsequent work.

As a young student at Harvard I developed two great interests. One was the fungi and other lower plants which were the province of my professor, William H. Weston. He was a charismatic teacher who exuded excitement for the possibility that lower (cryptogamic) plants, such as algae and fungi, made ideal subjects for experimental studies. He had many distinguished students, and when I started with him as an undergraduate, one of his finishing graduate students was John Raper, who was making the pioneer discovery of sex hormones in the water mold *Achlya*. Ralph Emerson had made similar significant advances with another water mold, *Allomyces*. While surrounded by these older students and Weston himself, I knew I wanted to be a cryptogamic botanist. But then I took a course in animal embryology with Professor Leigh Hoadley and was suddenly confronted with all the wonderful work of Hans Driesch, Wilhelm Roux, Hans Spemann, Edwin Grant Conklin, Ross Harrison, and many others who had advanced experimental embryology in the nineteenth and early twentieth centuries. I became entrapped all over again—I wanted to become two people. Then one day it dawned on me: why not work on the embryology (or developmental biology as it became to be known later) of lower plants!

To do this I had to find the ideal organism. Because I was surrounded by water mold enthusiasts it was very tempting to choose their area of interest. However, all that changed when one day I found the Ph.D. thesis of Kenneth Raper's (John's older brother), who had done his graduate work with Professor Weston a few years earlier. It described his discovery of *D. discoideum* and those wonderful experiments I mentioned. Here was exactly what I was looking for: the ideal non-animal embryo. I immediately wrote to Kenneth Raper, and he sent me cultures with some gracious encouragement that has kept me going for almost seventy years.

The question of how an egg develops into an adult was a matter of wonder and concern going back to Aristotle, and it blossomed at the end of the nineteenth century and into the twentieth: one could experiment on embryos and discover the causes of the developmental steps—why one stage produced the next. In the 1940s and 1950s it was realized that it was not just the embryos of higher animals and plants that developed, but development was a property of all organisms, from fungi to algae and other lower forms. Now the cellular slime molds showed themselves to be ideally suited for the study of experimental developmental biology. So first, following the tradition of Raper, an interest began not only in the description of the phases of their development, but in experimental studies parallel to those of causal embryology.

At first there was a difficulty because there was no known sexual system for the cellular slime molds. Later, when it was discovered, it turned out to be intractable and did not allow any way to do simple crossing experiments, unlike Mendel's model organism, the garden pea. In the beginning of the attack there were moderately successful ways of getting around this difficulty, and later, with the arrival of more and more clever molecular techniques, the disadvantage of the lack of ability to do crossing experiments virtually disappeared. The molecular genetics of the developmental biology of *D. discoideum* (which was now simply called *Dictyostelium*, reflecting its newfound status as a model organism) became central. The most recent high point in this program has been the sequencing of its entire genome: now it is possible to find out how many genes we share with a slime mold. The result is that our insights and understanding of the development of *Dictyostelium* have vastly increased.

In this joyous molecular roller-coaster ride there are many things about these slime molds that have to some degree been neglected, therefore stimulating me to write this book. Besides the central role of their development, many other aspects of their biology are equally fascinating, and here I would like to give the whole picture. As we shall see, they have not been ignored but simply overshadowed by the number of workers and publications in molecular developmental biology. Here I want to give all aspects of cellular slime mold biology equal time.

This means I want to give something closer to equal time to their evolution, their ecology, and their behavior, as well as their development. Within these big categories, starting with evolution, I will include discussions of their history on Earth, the taxonomy of the whole group and how they are related to one another, the origin of their multicellularity, and the interesting aspects of their peculiar sociobiology. The discussion of ecology will also involve their distribution in the soil, which is their natural habitat, their geographic distribution, and their mechanisms of dispersal. Concerning their behavior, besides the mechanism of how individual amoebae come together in the aggregation chemotaxis, two other matters are of major interest: one is the mechanism of locomotion of the migrating slugs, and the other is how slugs orient—what is the mechanism that makes them turn in towards light and sense heat and chemical gradients. More work has been done on aggregation chemotaxis than on any other aspect of slime mold biology; there is a rich literature. Turning to development, one aspect that has been studied in detail is the mechanism of differentiation of stalk cells and spores and how their proportions are controlled. Also there is the interesting question of the ways in which different species differ in their patterns of differentiation. As in all of biology, comparative studies showing differences among species are often helpful for a better understanding of the basic mechanisms; with all its advantages, there is a danger of clinging exclusively to one model organism. Finally, considering the great volume of work these days on the molecular analysis of development, I want to explore in what ways it has shed light on fundamental issues. It has done so, but some of the successes stand out as particularly significant.

2 THE LIFE CYCLE

One of the most striking things about cellular slime molds is their life cycle; it is so different. I remember the times when I first gave lectures on my experiments in the 1940s, and at question time the audiences of biologists seemed to show no interest in all the clever things I thought I had done, but they wanted to know more of what was then an unheard-of life cycle. Now that cycle can be found in every elementary biology textbook.

All the organisms with which we are most familiar, both animal and plant, including ourselves, start as a single cell, usually a fertilized egg, that takes in nutriment from a yolk (or a yolklike substance as in the cotyledons of a higher plant), and the egg begins to grow through repeated cell divisions, in that way achieving a large, multicellular state. In the case of animals, by the time the yolk is used up the embryo must have some other means of taking in energy: it must eat. This means that the yolk must sustain it long enough to allow time for the construction of a gut with a mouth to take in food and a tube that not only produces enzymes to break down the larger molecules to smaller sugars and amino acids, but also makes it possible for them to be absorbed through the gut wall. Animal alimentary systems are complex, and their invention must have had a long evolutionary history. Once achieved, it clearly was not reconstructed *de novo* for each animal group; all used the same basic food-processing machinery and the variations are later adaptations for the kind of food taken in. Some time ago, John R. Baker[1] wrote a short note in *Nature* pointing out that small photosynthetic organisms, compared to primitive animals, produced a great variety of shapes simply because they did not need a feeding device; they could take in all the energy they required and be any old shape: all they had to do is catch the Sun's rays. However, in both animals and plants they become large by growth made possible by a constant influx of energy.

The cellular slime molds acquire their energy in an entirely different way. They feed first as independent soil amoebae. Each individual amoeba surrounds a bacterium with its pseudopods, encases it in a food vacuole, and extracts the needed nutrients. Once they have cleaned an area of bacteria, they then come together; they aggregate to form a multicellular organism. Unlike animals and plants, they eat first; they grow by simply producing an increasing number of separate amoebae, and then when all the food is gone they stream together to become multicellular. It means that once they form their fruiting bodies they can no longer do anything that requires an intake of energy: they are static. The only part of them that is alive are the dormant spores. So we do have certain advantages in our kind of life cycle: we can keep going and enjoy life; we are not permanently frozen

[1] Baker, J. R. (1948).

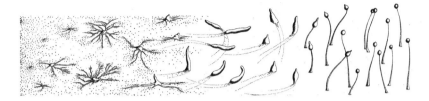

Fig. 1. The life cycle of *Dictyostelium discoideum* from the feeding stage (*left*), through aggregation, migration, and the final fruiting (*right*). (Drawing by Patricia Collins, *Scientific American* 1969)

once mature. (It should be noted that in general annual vascular plants do something very similar to slime molds. For example, a wheat plant will shoot up into the air in late summer, produce a bunch of dormant seeds at the top, and then the rest of the plant dies, turning a beautiful golden yellow.)

There is another group of organisms that do their feeding as single cells. Sponges are lined with flagellated chambers that sweep in currents of water from their submerged environment, and each of the flagellated cells will feed on incoming food particles. The difference between sponges and cellular slime molds is that sponges develop and eat at the same time, while slime molds separate the two processes in time and do one after the other.

Let me now describe in some detail the life cycle of one species of slime mold. I will choose the most familiar, *Dictyostelium discoideum*, but all are essentially similar (fig. 1). They normally grow in the soil where they are impossible to see, and therefore in the laboratory we cultivate them on a clear agar gel surface making it possible to follow all the steps of their progress as though they were living in a glass house. First they are supplied with bacterial fodder. This can be done by placing a loop-full of bacteria (almost any species of bacterium will work; *E. coli* is frequently used) on plain agar or agar with mild additional nutrients to encourage the growth of the bacterium. Then a few slime mold spores are added, and, in a matter of hours, they will split open and from each spore emerges a single amoeba that immediately begins to feed on the surrounding bacteria. As they fatten they will, at about three-hour intervals, divide in two so that before very long there will be vast numbers of them. The amoebae themselves are quite small, about the size of our white blood cells, and in fact they look rather similar.

The stimulus to become multicellular has something to do with the sudden absence of the food. The bacteria have all been eaten. In a loose way this is often referred to as starvation, but half-digested bacteria can still be seen in the food vacuoles of the amoebae. In any event, their behavior changes and they suddenly zip about much faster, something that can be seen dramatically in time-lapse movies.[2] This phase lasts a number of hours, and then aggregation shows the first signs of starting. A few cells, or small groups of cells, start emitting an attractant, and the amoebae around those hot spots turn and orient towards the source. Before it was known what was the chemical nature of the attractant, it was called *acrasin,* inspired by a literary allusion to Edmund Spenser's *The Faerie Queene.* These slime molds are members of the *Acrasiales,* and there is a witch in *The Faerie Queene* who, like Circe, attracts men and turns them into beasts. The parallel seemed apt because acrasin not only attracts cells, but much later it was discovered that it also plays a role in transforming

[2] Samuel (1961) made careful measurements of the speed of the amoebae during development.

them into spores and stalk cells. Today we now know that the chemical nature of the acrasin for this species is cyclic adenosine mono phosphate (cyclic AMP), a substance found everywhere and important in animals for passing on the signals from hormones into the cells.

As the cyclic AMP is pumped from the newly formed centers, it has two effects on the neighboring amoebae: it attracts them so they not only orient towards the centers, but induces those surrounding amoebae to start emitting cyclic AMP on their own. They in turn become emitters so the attraction wave spreads outward, causing a myriad of amoebae to start streaming towards the center. In contrast to the feeding amoebae, they now are elongate and polarized. By this I mean their front ends appear more obviously different from their hind ends. They also become sticky and tend to attach to one another. As they come together in chains, the front end of one amoeba will stick to the hind end of another, and these chains coalesce to form broad streams of incoming amoebae. If a single amoeba is attracted to such a stream (which is pumping out cyclic AMP), it will insert itself between two cells in the stream so that its front end attaches to the rear of the anterior amoeba and its posterior end to the front of the amoeba behind it (fig. 2).

Usually during aggregation the cyclic AMP is given off in pulses so that the incoming amoebae show waves of fast movement towards the center. It is a relay which is the result of cyclic AMP stimulating the production and secretion of cyclic AMP in the amoebae downstream. Again, these waves can be seen dramatically in time-lapse movies and have been intensively studied. Although this is the norm, it is also well known that amoebae can equally well aggregate in a steady gradient of attractant.

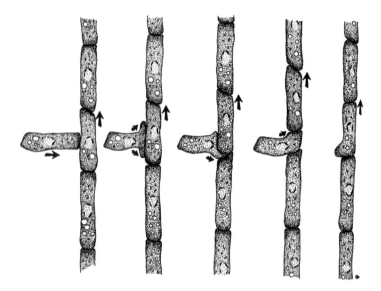

Fig. 2. Drawing of an amoeba entering an aggregating stream. Note that it achieves an end-to-end attachment. (Redrawn by R. Gillmor from Shaffer 1961)

The aggregate center is a mass of amoebae surrounded by a thin, transparent slime sheath (hence the name "slime mold"). It soon forms an elongated migrating slug that looks rather like a finger; it has a distinct front end and a hind end and moves forward at the sedate pace of about a millimeter an hour. Since growth occurred at the single-cell stage, its size depends on how many amoebae have entered the aggregate, and slugs will vary in length from about 0.2 to 2 millimeters, a tenfold range, and by the latest estimates the number of amoebae they contain ranges from about 10,000 to 2 million. As we shall see later, this migration serves the purpose of bringing the amoebae from a feeding area in the soil where the bacterial food was found to an area ideal for fruiting, namely on top of the soil where the spores can be dispersed by passing insects and other invertebrates.

Once the slug reaches the soil surface it eventually ceases its forward movement and points up into the air. Already in the migrating slug there are early signs of differentiation: the anterior fraction of the amoebae are to become stalk cells, and the larger proportion of posterior amoebae will become spores. The final differentiation of these two cell types occurs during fruiting. First, stalk cells, which are large and show the growing presence of vacuoles, appear as a group near the tip. Then, as they enlarge they begin forming what eventually will be a cylindrical, cellulose stalk sheath. The amoebae that surround the sheath at the top enter the stalk-forming zone and they too start transforming into stalk cells. The result is a cellulose cylinder filled with vacuolated cells that is slowly pushed down the middle of the cell mass by the amoebae moving in and pressing at the top. During the course of this process the vacuoles in the cells keep enlarging, and in fact are dying: they are like the pith of a vascular plant. As the thick cylinder of cellulose is deposited around these cells, cellulose is also deposited along the outer surfaces of the cells. The final result is a stiff tube with supporting cross-struts—an engineer's delight that is admirably suited to lift and hold the spores up in the air. Those cells that keep piling on top of the elongating stalk—the prestalk cells—eventually become depleted, and when they do the fruiting body has ended its rise up into the air. This species has one additional feature: there is a group of cells around the bottom of the stalk that also becomes vacuolated, forming a pedestal, a basal disc; hence the species name, *discoideum*. While rising up into the air, each of the posterior amoebae,

which will become spores, encapsulates into an elliptical cellulose shell, and as a group they are lifted up at the tip as the
stalk rises.

This is not the only cycle in cellular slime molds: there
also is a sexual cycle. If the amoebae of two mating types
come into contact, and the conditions are right, there will be
a fusion of two amoebae to form a zygote. This is accompa-
nied by a normal aggregation, and the amoebae will cluster
around the zygote. Ultimately the cell mass becomes spheri-
cal and surrounded by a cell wall. The zygote then proceeds
to cannibalize the other amoebae while it undergoes meiosis.
Unfortunately it has proved very difficult to germinate the
macrocysts in the laboratory, making genetic crossing experiments impossible. It is not known how common the sexual
cycle is in nature, but one might guess that it is relatively
rare, especially compared to the asexual cycle.

Other Species

If we look at species of cellular slime molds other than *D. discoideum*, we will find numerous different phenotypes (fig. 3).
Most species do not have a stalkless migration stage, as for
instance *D. mucoroides*, which continuously forms a stalk from
the end of aggregation on; first it moves horizontally, and then
eventually, when it has reached the surface of the soil, it may
move up vertically. A number of species have multiple spore
masses (sorus; pl. sori) instead, and they exist in a variety of
configurations; most common are the two species of *Polysphondylium* that form a series of evenly spaced whorls, each
bearing a minute terminal sorus. The spores, and therefore
the sori, can vary in color, from white, to yellowish, to purple,

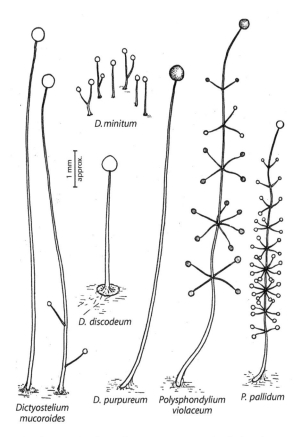

D. minitum

1 mm approx.

D. discodeum

Dictyostelium
mucoroides

D. purpureum

Polysphondylium
violaceum

P. pallidum

Fig. 3. The fruiting bodies of various species of cellular slime molds. (From Raper 1941)

as in *D. purpurium*. Most species have no *D. discoideum*-like basal disc, but only the unadorned end of the stalk touching the substratum. A few species have a structure at the stalk base called a crampon that looks like a grasping hand. There is an interesting exception to the norm: the stalk of *Acytostelium* is not made up of dead vacuolate cells, but it is totally cell-free: all the amoebae secrete a delicate cellulose stalk, and then those same cells all turn into spores. Species vary in size, from the large *D. giganteum* to the small *D. minutum*.

Years ago, during the Soviet days, I was asked by my university to talk to two presidents from agricultural colleges in Russia who happened to be visiting. I had no clue what to talk to them about, so in desperation I started telling them about my research. Having to speak to them through an interpreter was no help, and both looked desperately bored as I described the slime mold life cycle: what on earth was I talking about? With a sudden inspired flash I wrote on the blackboard "social amoebae" and immediately they sat up straight, smiled, and began asking questions to learn more. It is the only case I know of where the life cycle of a cellular slime mold played a role in international relations.

3 EVOLUTION

Cellular Slime Mold Origins

We want to know how and why amoebae became social in the first place. This key question revolves around a second question: what advantage does togetherness provide them, or, to put it in a better form, how could it have been encouraged by natural selection?

To begin, becoming multicellular means an increase in size, and this is directly related to the key matter of dispersal. Organisms that have an effective mechanism for spreading, for dispersing, can more easily find new sources of food and multiply. Natural selection involves competition—a struggle for existence—and the victors are the ones that produce the most progeny. If at one time in Earth history all the soil amoebae were single and a more effective method of dispersal arose by chance, those that possessed it would flourish. Numerous species of soil amoebae have cysts, where each individual amoeba becomes encased with a hard cell wall, capable of surviving periods of drought or other adverse conditions. The question then becomes, what is the most effective way to disperse these resistant bodies to improve their chances of falling on some food-rich territory?

We must first examine how dispersal is effected in cellular slime molds that exist today. There is considerable evidence that the prime method involves insects and other small invertebrates touching the fruiting bodies, picking up the sticky spores on their appendages, and then brushing them off at distant places where they have wandered. The minute fruiting bodies sticking up into the air seem admirably constructed for this purpose. I have seen the phenomenon when I bring some unsterilized forest soil into the laboratory and place it in a covered glass dish a couple of inches high. If the conditions are right for cellular slime molds to flourish, they will be abundant on the soil surface. And should there be, by chance, an ant or a beetle larva, or a small spider, or a mite, or a millipede in the soil sample, in no time at all the slime mold stalk tips will be barren—the spores have been carried off by a passing beast.

Let us turn to the key matter of when, during the course of evolution, solitary amoebae became social. If fruiting bodies are specifically adapted to being dispersed by passing soil animals, they could not have appeared on the earth before the soil had been conquered by those animals. We must therefore ask when was the Earth first crawling with small soil animals, which, if my argument is correct, must have preceded the origin of the multicellular slime molds.

The conquest of land by small animals probably occurred some time in the Ordovician, well over 400 million years ago—although this is an active subject of inquiry in paleontology, and it may well be pushed back to an even earlier time.[3] Of course, solitary amoebae are no doubt vastly older; we are only talking about the time they became multicellular.

The Cellular Slime Mold Family Tree

Around 1900 it was thought that there were about seven species of cellular slime molds; when I began to work with them in the early 1940s about ten were recognized; today, through the discoveries of a number of workers, about one hundred species have been identified. Often the variations in their morphology are minor, but they can be pigeon-holed as distinct species. As with many microorganisms, the question of what constitutes a species and how the concept of species compares to the well-established and accepted one for higher animals and plants is a muddy matter. The tradition for cellular slime molds' classification is entirely based on their morphology. I can remember way back discussing this point with Kenneth Raper, who himself discovered many of the new species, and he was quite adamant that the classification of the group was for the purpose of making them easy to identify; it said nothing about their phylogenetic relations. This is in the spirit of Linnaeus, who thought each species was created by an act of God, and who has been the basis of taxonomic keys beloved by some (not me!) through the centuries.

The idea that it might be possible to study the evolutionary relation between species has come into bloom only recently. It is of great importance to unravel the ancestry of any living organism, and now, with new molecular methods, we

[3] The Ordovician spans from 443 to 488 million years ago. I would like to thank Greg Retallack and Gerta Keller for help with the question of when soil with crawling animals might first have appeared in Earth history.

can do this with amazing detail and accuracy, and as we will see this has been done for the slime molds. However significant such studies have turned out to be, I agree with Raper that we still need a field guide to the cellular slime molds and should keep the old system of morphological classification alongside the molecular phylogenetic tree. Both are needed.

Such a molecular phylogenetic tree has just been published by Sandie Baldauf and Pauline Schaap and their numerous collaborators, including two leaders in cellular slime mold taxonomy, James Cavender and Hiromitsu Hagiwara.[4] They have looked at two genes that tend to be stable for long periods of time (conserved) and followed their variations for seventy-five species. With this tremendous wealth of information they have laboriously built a family tree that tells us a lot about both the relationship of the various species and their evolution (fig. 4).

As can be seen in figure 4, they separate out into four major groups (or clades), group 1 being the most ancient and group 4 the most recently evolved. There are many interesting things in this grouping, but the most surprising one to me is that the two common whorled *Polysphondylium* species turn out not to be closely related to one another, which I—along with everyone else—had assumed to be the case all these years. They look so much alike, each with their beautiful, symmetrical whorls that jut out like spokes of a wagon wheel. Furthermore, they have some other unique features: both have the same acrasin, the dipeptide glorin,[5] and unlike any other known species both have single "founder cells" that initiate aggregation; in other species the beginning of

[4] Schaap et al. (2006).
[5] Shimomura et al. (1982).

aggregation is started by a group of amoebae.[6] The main obvious difference between them is that *pallidum* has white spores, and *violaceum* has purple ones. This would imply that their elegant architecture and those other features have been invented independently twice at different times, perhaps many, many years apart. One caution should be mentioned: the relation between these two species is based on only two genes, and it will be important to examine their differences involving more genes.

Also, the authors point out that the more modern group 4 tend to be unbranched with single sori at their stalk tip. Branches have gone out of fashion in later slime mold evolution, but that has made them neither more nor less successful than the branched ones in reaching the twenty-first century.

Another interesting trend is revealed by this family tree: members of the most modern group 4 tend to be larger than the more ancient species. Although I am inordinately fond of organisms getting bigger during the course of evolution, I am uncertain about the significance of this trend in the slime molds. Part of my problem is that the size difference is modest, but more importantly the large and small species all coexist today. The larger species do have one obvious advantage in that they are able to migrate greater distances and therefore have a better chance of reaching the surface of the soil from the depths below. Yet the smaller species not only continue to survive, but they are more ancient and have thrived for millions of years, many of them in the presence of the larger (group 4) species. Size increase is a major advance during the course of evolution in the step from unicellular amoebae to

[6] Shaffer (1961).

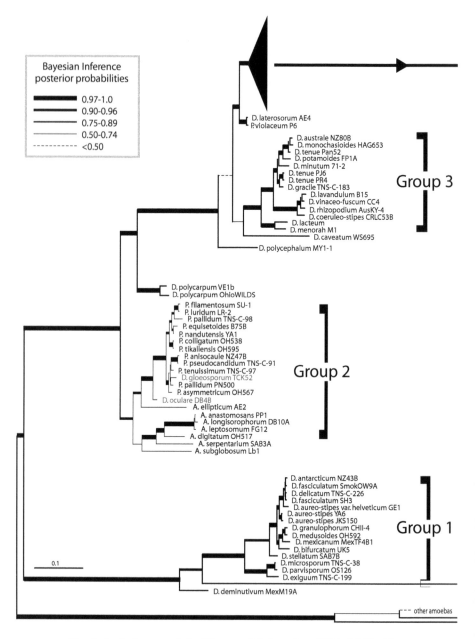

Fig. 4. A phylogeny of the Dictyostelids based on SSU rDNA. (From Schaap et al. 2006. Reprinted with permission of the AAAS)

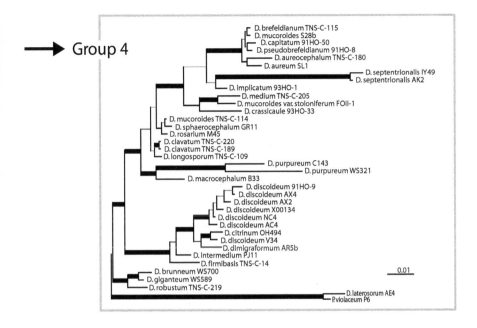

Group 4

D. brefeldianum TNS-C-115
D. mucoroides S28b
D. capitatum 91HO-50
D. pseudobrefeldianum 91HO-8
D. aureocephalum TNS-C-180
D. aureum SL1
D. septentrionalis IY49
D. septentrionalis AK2
D. implicatum 93HO-1
D. medium TNS-C-205
D. mucoroides var. stoloniferum FOII-1
D. crassicaule 93HO-33
D. mucoroides TNS-C-114
D. sphaerocephalum GR11
D. rosarium M45
D. clavatum TNS-C-220
D. clavatum TNS-C-189
D. longosporum TNS-C-109
D. purpureum C143
D. purpureum WS321
D. macrocephalum B33
D. discoideum 91HO-9
D. discoideum AX4
D. discoideum AX2
D. discoideum X00134
D. discoideum NC4
D. discoideum AC4
D. citrinum OH494
D. discoideum V34
D. dimigraformum AR5b
D. intermedium PJ11
D. firmibasis TNS-C-14
D. brunneum WS700
D. giganteum WS589
D. robustum TNS-C-219
D. laterosorum AE4
P. violaceum P6

0.01

D. multi-stipes UK26b

Dermamoeba algensis

Thecamoeba similis

the formation of fruiting bodies made up great numbers of amoebae, achieved through natural selection for more effective dispersal.

Sociobiology of Cellular Slime Molds

One of the interesting consequences of the peculiar life cycle of the cellular slime molds is that, unlike organisms that become larger by growth, it is far easier for them to have a mixture of cells that might be related to different degrees; in other words, to form chimaeras. After all, presumably any compatible amoeba could enter an aggregate and become part of a fruiting body, while in animals or plants they might acquire genetically different cells by mutation and the growth and multiplication of that mutant cell. (This is not quite true, for as Leo Buss has pointed out there are some invertebrates, some fungi, and a few plants where fusion of different cell lineages occurs.)[7]

The classic first study of the phenomenon in slime molds goes way back to a paper of K. B. Raper and C. Thom, who addressed the question as to whether two species could aggregate together and form hybrids.[8] Their answer was they could not, but the ways of preventing the intermixing of the amoebae were especially interesting. If they mixed about an equal number of amoebae of *D. mucoroides* and *P. violaceum* on an agar culture dish, as they aggregated there would not only be no mixing at all, but the streams of the aggregates of the two species climb over one another: they seemed totally unaware of each other's presence. Today we know that those

[7] Buss (1982).
[8] Raper and Thom (1941).

particular species have different acrasins: the chemoattractant for *D. mucoroides* is cyclic AMP, while that of *P. violaceum* is the dipeptide called glorin. (At the time of their experiments it was not even known that aggregation consisted of a chemical attraction of the amoebae to the centers.) This means that when the aggregation centers are first formed, they are producing their own acrasin and will attract only the amoebae that respond to it; they will have no interest in the attractant of the other species and therefore no possibility of commingling.

In their second case, Raper and Thom chose two species that had the same acrasin, which is cyclic AMP. They were *D. mucoroides* with white sori, and *D. purpureum* with purple sori. When they did the parallel experiment, these two species co-aggregated into common centers, but there was a surprising sequel. Fruiting bodies arose from the same mound and their sori were either white or purple: the amoebae had separated into two groups in the mound, and the resulting fruiting bodies were pure and all their amoebae were of either one species or the other.

The amoebae of the two species that are moving around in the mound will tend to stick more closely to their own kind; there is something about their surface chemistry that allows them to recognize kin and adhere more strongly to one another than to the cell of a foreigner. This has also been shown in a different way. If one artificially mixes the amoebae of two species, or two strains of one species by grafting the tip of a slug of A onto a decapitated late aggregation of B, the amoebae will never intermingle, but, again, kind will stick to kind.[9]

[9] Bonner and Adams (1958).

What is most remarkable is that in one combination if the amoebae of a fast-moving species are put behind those of a slow-moving one, as the slug crawls forward the faster amoebae will percolate through the slower ones, and once they arrive in front they will form a homogeneous clump of cells. Regardless of the disturbance, the same species literally stick together.

Before examining the broader significance of the behavior of genetically distinct amoebae in a cell mass, let us first address the question of what happens in nature. Because of their curious life cycle, cellular slime molds could be made up of identical cells all arising from a single spore, that is, a clone, or they could be made up of genetically different strains of amoebae that just happened to be in the sphere of influence of a center emitting their acrasin. What do we find in nature?

A number of studies have shown that, even in a very small sample of soil, if separate amoebae are isolated and cloned and then tested for their DNA fingerprint (slime mold forensic DNA testing!), genetically diverse strains of the same species will cohabit small domains of soil; there is genetic diversity in nature.[10] This means, and has been demonstrated, that the fruiting bodies in nature can be chimaeras. In fact, this could be a common condition, although clones are probably common as well.

One of the most recent and comprehensive studies (which cites all the earlier ones) is on the species *D. giganteum* from a nature preserve in India where V. Nanjundiah and his colleagues found just such a genetic diversity.[11] They brought

[10] Reviewed in Kaushik and Nanjundiah (2003); see their p. 842.
[11] Kaushik, B. Katoch, and Nanjundiah (2006).

the different strains to the laboratory, some being close neighbors in the forest soil, and artificially manufactured chimaeras. Here they also found, using suitable molecular markers, that the amoebae of one strain tended to clump together within the chimaera.

The big question that they and others have addressed is, what are the differences between clonal and chimaeric fruiting bodies? It can be answered at two levels: that of collective, group selection and that of the level of the individual amoebae. As far as the group is concerned, are there any advantages or disadvantages for the fruiting body as a whole if it is made up of one or more genetic strains? The balance of the evidence from Kaushik and Nanjundiah is that there is no significant difference in the success of the two types of fruiting bodies; they are both equally capable at producing spores. However, the interesting matters arise when we look at the struggles of the individual amoebae, and for this I will go back to much earlier studies.

The first hint came from the work of Michael Filosa in which he examined an old culture of *D. mucoroides* (which had originally been isolated from some giraffe dung from the zoo—a curious but irrelevant fact) and had been recultured continuously in the laboratory over some years, always by reinoculation with many spores at a time.[12] He wondered if it was a pure clone, or was it a composite of genotypes? By isolating individual spores he produced clones and found that indeed the isolates were not identical: the stock culture was chimaeric. One of the strains when cloned was very abnormal and produced a remnant of a stalk but normal spores.

---

[12] Filosa (1962).

If he mixed this strain with one that gave normal fruiting bodies, the mixture always produced fruiting bodies of a normal appearance. (This was also true of a strain that was incapable of aggregation; it too behaved normally if mixed with a normal strain. He did not use the word "chimaera" because that was long before it became fashionable, nor the word "cheater," another word that has come into the current literature. The strain was not doing its duty in producing the altruistic stalk. In fruiting bodies invariably all the stalk cells die in the process of raising the spores into the air; they are the ultimate altruists. When the stalk-deficient strain formed chimaeras with a normal strain, they could benefit from a free ride on the normal stalk.

The phenomenon was greatly extended by Leo Buss who took soil samples just millimeters apart, cloned the amoebae from the samples, and found that different strains of *D. mucoroides* were living in close proximity.[13] This was also shown to be the case for *D. giganteum* by Kaushik and Nanjundiah. Buss found that one of the strains was an extreme cheater: it produced only spores when grown alone. Again, when it was mixed with a normal strain the mixture produced a normal fruiting body. The cheater was taking that free ride on the stalk of the other strain. He pointed out that this mutant was an auto-parasite, which is another name for a cheater.

A number of recent studies have addressed the question of whether the amoebae within a chimaera affect one another. Do the amoebae of Buss's mutant force its host to make more stalk cells and fewer spores as would seem to be the case? There is clear evidence that in some instances the amoebae of

---

[13] Buss (1982).

the mixed strains do affect one another, and some of the examples are indeed curious. For instance, it was shown by A. J. Kahn that a mutant of *D. purpureum* that was incapable of aggregation, when mixed with the wild type, not only became incorporated into the normal fruiting body, but it became permanently cured and subsequently produced normal fruiting bodies all on its own.[14] I do not know quite what to make of this remarkable transformation, but it certainly shows communication between the strains. In other cases, as suggested above, a defective mutant can persuade the wild type to produce more stalk cells.[15]

To me the most remarkable and interesting example was discovered by Filosa (described in detail in his doctoral thesis, which was the basis of his original paper). His laboratory strain of *D. mucoroides* was a mixture of strains, and one of them produced just vestiges of a stalk and therefore a cheater. When he first looked at the stock culture he found this mutant to comprise about 9 percent of the spores, and remarkably this ratio remained roughly constant for twenty-five generations. He could even start with a different ratio and within a few generations the ratio reverted to the approximate 9 percent.[16] It is like a parasite that does not destroy its host but sees to it that the host lives on. Such a state is obviously to the advantage of the cheater, the parasite; it is the way its existence can be maintained by natural selection. For the slime mold it is a sophisticated exchange between the

[14] Kahn (1964).
[15] This was elegantly shown by Ennis et al. (2000) using molecular techniques. By mutagenesis they produced a cheater mutant and demonstrated in detail that it induced the host to produce more stalk cells while it produced none.
[16] This was also shown by Buss (1982).

strains that allows long-term stability. Unfortunately, we do not know the mechanism, but it must be very similar to the balance between host and parasite. And it could not be a clearer case of communication between the cell strains within a slime mold chimaera.

It should be added that recently it has been shown that cheaters can maintain themselves by another tactic. By artificially inducing mutants in a laboratory strain of *D. discoideum,* Santorelli et al. (2008) selected for cheaters by mixing the mutants with the parent strain. If, in the mixtures, the mutants produced an abnormal number of spores, they were considered cheaters. The interesting thing is that these cheaters were cryptic, because when cloned so that they were 100 percent mutant amoebae, they produced the normal number of spores. They only cheat when mixed with the parent strain. It would be important to know if these interesting cheaters exist in nature.

CHIMAERAS, HETEROKARYONS, AND HETEROCYTONS

There is one other point to be made about chimaeras that was discussed by J.B.S. Haldane for some fungi.[17] Filaments, or hyphae, of ascomycetes have their nuclei wandering about freely within the tubes. These hyphae will regularly fuse and join with one another, and as a result will have nuclei from different genetic strains within one hyphal network. This is called a *heterokaryon*, and it will produce haploid spores that will differ genetically. After they germinate they will fuse again with other strains to form a new heterokaryon. Such a

---

[17] Haldane (1955); see his p. 20.

system is capable of producing variation in the offspring, and as Haldane points out, has the advantage over sexuality in that there can be more than two parents; but it lacks the advantage of chromosomal recombination that can result from the formation of diploid nuclei and the subsequent meiosis. The phenotypes of the offspring are chimaeric and as a result vary depending on what combination of genotypes they possess. The real difficulty is that it is just as easy to undo an advance as to create one, so it is hard to see how this system will produce a stable, major evolutionary step.

What happens in cellular slime molds is very similar. The main difference is that each slime mold haploid nucleus is surrounded by a cell membrane; they are true cells. For this reason it has been called a *heterocyton*, in contrast to a heterokaryon.[18] The phenotype of such a chimaera or heterocyton is the result of the combination of signals from its collective genotypes. Furthermore, when the individual spores are scattered by a passing insect, they can, by coming together in different combinations, form a variant chimaera or even a clone. As we have seen, most cellular slime mold heterocytons have the morphology of the wild type and the mutant phenotype is suppressed. This is undoubtedly because the selection is at the cell level and the mutants are cheaters; they have not produced a successful phenotypic morphology of their own, but gain through their parasitic activities on the effective phenotype of the host. For these reasons it is hard to imagine that their ability to form chimaeras has led to any great evolutionary step forward, such as the formation of a new species. Rather, it seems to be a minor struggle

[18] Filosa (1962).

at the cell level. However, it must be remembered that slime molds have been forming these chimaeras for millions of years, and it is not inconceivable that some novelty might have come from it.

To summarize, the fact that cellular slime molds grow first and then become multicellular does offer them the opportunity to form chimaeras, which, indeed they readily do. The assembled genetically different amoebae do communicate and affect one another in the cell mass; they are even able to order and control the fate of their cohabiters. This means that at the cell level there is great activity: communication, competition, suppression: it is an ideal sociobiological playground. And indeed various laboratories and theorists, such as Joan Strassmann, David Queller, Richard Kessin, their co-workers and others have taken full advantage of it.[19] However, it is difficult to see how all these activities at the level of the cells could lead to the formation of a new species.

Let us now dwell on major evolutionary changes. We have seen that with mixed genomes an individual multicellular slime mold retains pretty much the same phenotype no matter what its internal cellular mix. The question is, how do different-shaped fruiting bodies, different species arise? Even if a particular combination of cells in the chimaera would produce a new morphology or a change in the physiology, it might be lost in the next generation. Many species are known to have macrocysts, which is a sexual stage, and although it might be an infrequent event, they could, through meiosis, incorporate a permanent genetic change that would be stable provided it was approved by natural selection. About

---

[19] Strassmann and Queller (2007); Dao et al. (2000)—in this paper there are further references to the work of others.

one hundred species are known today, and there have been many millions of years and an untold number of generations for them to have arisen.

SLIME MOLD AGGRESSION

Not all cellular slime molds are peaceful and cooperative: there are two very interesting examples of distinctly unfriendly aggression. Indeed, they are not the norm, and so far as we know they are rare, but that makes them no less interesting examples of their social behavior.

One is the discovery of H. Hagiwara of a strain of *P. pallidum* collected in Japan that produces a substance that destroys many other strains of *P. pallidum* as well as a common wild-type strain of *D. discoideum*.[20] It has not been tested on other species, but where it does work its effect is the dramatic killing of the amoeba of a vulnerable strain or species. They do so by secreting a lethal molecule (possibly a protein) that devastates the amoebae of the susceptible victim. It is hard to interpret the sociobiological significance of this phenomenon. There are similar cases among other microorganisms. It would seem to indicate intense competition, yet it is such a rare phenomenon that clearly it does not play a major role in the evolution of cellular slime molds.

In terms of mechanisms, the other example is quite different. David Waddell decided to see if slime molds found on bat dung deep in a cave might have some blind forms.[21] He failed to find species that did not respond to light, but he

---

[20] Hagiwara (1989); Mizutani et al. (1990). See also Mizutani and Yanisagawa (1990).

[21] Waddell (1982).

found something much more interesting. He discovered a new species that turned out to be a very busy carnivore (which he cleverly named *Dictyostelium caveatum*). Since slime molds feed as separate amoebae, this species attacks by entering the aggregate of another species, and the *caveatum* amoebae systematically eat—by engulfing—the amoebae of the victim species. It is a carnivore from within. Its effectiveness is quite remarkable. If the slug of *D. discoideum* has as few as one *caveatum* amoeba in ten thousand *discoideum* amoebae, after a period of what looks like normal migration it will suddenly stop, looking a bit piqued, and small fruiting bodies will sprout out all over its surface. These are all the fruiting bodies of *caveatum*; not one *discoideum* spore can be found. It reminds one of a parasitic wasp that will lay its eggs in a large caterpillar that will eventually die and sprout myriads of wasp progeny all over its surface.

Again it is hard to assess the evolutionary significance of this behavior, largely because we only have one case and assume therefore that it is rare. Perhaps it merely shows a good example of opportunism on the part of one species, something that has occurred in so many other groups of organisms. It is interesting that it invented this behavior in spite of its peculiar method of eating as individual cells.

4 ECOLOGY

Geographic Distribution of Cellular Slime Molds

There are some species of cellular slime molds that are found everywhere and others that seem to be restricted in their distribution. Our most comprehensive understanding of slime mold distribution comes from the work of James Cavender and his co-workers.[22] They show that a number of cosmopolitan species are found everywhere, other species are found in large geographic areas, such as the tropics or a major area of a continent, and still others are endemic and found on islands such as New Zealand.[23]

They also point out that cellular slime molds are primarily associated with forest soils, and they are to some extent affected by the kind of trees: some species thrive in deciduous forests, while others in coniferous ones. *D. rosarium* specializes in arid, chaparral habitats, while *D. sphaerocephalum* is found in northern climes in tundra (in contrast to the numerous species found only in the tropics).

[22] Swanson et al. (1999).
[23] Cavender and Stephenson (2002).

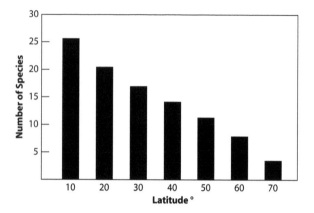

Fig. 5. The number of cellular slime mold species at different latitudes. (From Cavender 1973)

Latitudinal Diversity

Cavender makes another very interesting observation in an earlier study.[24] Far more morphological species of cellular slime molds are found in the tropics than in colder habitats; in fact there is a clear relationship between latitude and the number of species (fig. 5). Furthermore, more recent studies show a similar relationship in mountains where the number of species decreases the higher the altitude.[25] This is intriguing because it exactly mirrors what is well known in larger organisms, as for instance in vascular plants, birds, and mammals. That species richness increases with warmer temperatures is a universal phenomenon in both the macro and micro world.

[24] Cavender (1973).
[25] For a brief review and references see Raper (1984), p. 42.

There is a vigorous debate as to why large animals and plants become more diversified in warmer climes. The traditional view, reflected in the writings of Evelyn Hutchinson, is that the driving force is the diversification of niches: the more separate cubby holes in the environment, the more species to occupy them. There is another view championed by Stephen Hubbell: his neutral theory of species abundance.[26] He argues that the diversity can be understood in terms of stochastic processes such as immigration, emigration, extinction, along with the formation of new species, in keeping with the ideas of R. MacArthur and E. O. Wilson based on their study of island biogeography.[27] It is generally agreed today that both processes can be involved and they are not mutually exclusive.

Undoubtedly this second mechanism is dominant for the cellular slime molds. It is generally assumed that for larger plants and animals, a number of factors contribute to this gradient in diversity, but only one of them seems to apply to the same trends in cellular slime molds, which is climatic stability. In the tropics and in the valleys between the mountains, the climate is relatively steady and unchanging, while as one moves away from these zones, one finds the season's temperatures fluctuate with increasing intensity.

For slime molds these annual changes will cause the destruction of the amoebae and spores in the wintertime (and summertime, due to the heat and the dryness). This has been shown by sampling soil in temperate zones at different seasons and finding a loss of slime molds during winter and

[26] Hubbell (2001).
[27] MacArthur and Wilson (1967).

summer.[28] In other words, for most species the only way to replace those lost by seasonal extremes is by dispersal, and this reseeding plays an increasingly crucial role the harsher the seasonal environment.

Dispersal

The fruiting body is the prime agent for spore dispersal, so important to reach another patch of bacteria for continued survival. Spore dispersal of this kind is a widespread phenomenon. If one examines the surface of soil in a moist chamber, one cannot but be impressed by the number of small fruiting bodies that appear on the surface. Almost invariably there will be some species of fungus, for there are a vast number of species of mold that have filaments or hyphae that rise to hold small terminal masses of spores. For instance, one finds many common species of phycomycetes, such as those of *Mucor*, or species of the innumerable *fungi imperfecti*, such as those of *Aspergillus* or *Penicillium*, and a vast number of others. This has been demonstrated recently in a survey of forest soil in India where a few hundred different species of molds are found in the same locale.[29] Small fruiting bodies present the most massive collection of examples of convergent evolution among all of living organisms; these minute structures have been reinvented over and over again in totally disparate groups—not only in slime molds and filamentous fungi, but also in bacteria (the stalked, multicellular myxobacteria). There is even one example among ciliate protozoa. One can only

---

[28] See Raper (1984), p. 46.
[29] Satish et al. (2007).

conclude that the selection pressure for spore (or propagule) dispersal among small soil organisms must be intense, and the same solution has been independently rediscovered an endless number of times.

In slime molds, dispersal is known to occur at different levels—to extend to different distances. Some species have a dispersal mechanism at the separate-amoeba feeding stage. The individual amoebae of those species repel one another, and it is assumed that this is a mechanism for spreading the amoebae out evenly, favoring effective grazing on a lawn of bacteria.[30]

Of course, the important dispersal is that of spores to start a next generation. As mentioned earlier, there are many bits of evidence to favor the idea that invertebrates in the soil play a major role in carrying the spores away from the fruiting site. For instance, nematodes are known to eat the amoebae of cellular slime molds, but they often swallow spores, and these are not digested but pass through the gut and are spread by the motile worm.[31] The distances involved are very short; other soil animals are far more effective. Greater distances can be achieved by earthworms and pill bugs, for they too can pass spores unharmed through their gut.[32]

The spreading of spores by small animals can be seen in the laboratory in the soil blankets spread over an agar surface, as mentioned previously. In the sori the spores are held together by a sticky substance, and as soon as anything touches them, be it an ant leg (or an inoculation needle), the spores will stick to it and can disperse as far as the animal travels.

[30] Keating and Bonner (1977).
[31] Kessin et al. (1996).
[32] Huss (1989).

There can also be slime mold dispersal over far greater distances. Ground-feeding mammals such as rodents will accumulate spores in their dung that they will deposit some distance away.[33] V. Nanjundiah has informed me that his student Sonia Kaushik, who worked in an extensive nature preserve in central India, found slime mold spores in the dung of a number of large mammals: elephants, barking deer, sambar, panther, and hyena. Some of these species will wander many miles. However, the record goes to migrating birds. In an important paper, Hannah Suthers collected bird feces at a banding station and found an abundance of cellular slime mold spores.[34] She showed that spores could remain viable in a caged bird's gut in the laboratory for ten days. For a migrating bird this would mean that the slime mold spores could be carried great distances, for instance from North to South or Central America in the fall, and the reverse in the spring. This variety of dispersal distances is of interest to mathematicians, who call it "anomalous diffusion."

The propagules of many plants and fungi are carried by the wind, but not cellular slime molds. Since the spores stick to one another they cannot be blown apart. On the other hand, undoubtedly rain effectively spreads the spores: a deluge will certainly sweep them downstream. While rain can be effective once the fruiting bodies with their spores are mature, it will drown the earlier stages of their life cycle; they are not aquatic organisms. They have, however, invented a strategy to avoid the detrimental effects of a flood. Under water they can produce macrocysts that are also formed as a result of aggregation by chemotaxis and are known to exist

---

[33] Stephenson and Landholt (1992).
[34] Suthers (1985).

in a number of species of dictyostelids. As we saw, they are the sexual stage if two mating types are brought together. However, in many of the cases examined a single strain will form macrocysts without sexual fusion.[35] Macrocysts serve two functions: sexual recombination, and forming a resistant stage when they are inundated by rain and therefore unable to produce normal fruiting bodies and spores.

It should be added that many species of cellular slime molds produce microcysts, where separate amoebae that do not aggregate become encapsulated in a cellulose casing. This is another way to produce resistant bodies when the conditions are unfavorable for fruiting.

D. mucoroides, v. stoloniferum

This variant of *D. mucoroides* provides an interesting footnote to slime mold dispersal. The rule is for cellular slime molds to produce an inhibitor of spore germination. The adaptive rationalization of this phenomenon is that if a mass of spores lands in a colony of bacteria, the amoeba from just one germinating spore is sufficient to invade, multiply, and eat the bacteria. If the other spores are inhibited from germinating, they can be saved for starting a new generation elsewhere.

Stoloniferum does things differently: it has no inhibitor and all the spores in a group germinate.[36] This means that as soon as a spore head hits the substratum many amoebae will appear, and if there is no food about, they will immediately

[35] See chapter 9 in Raper (1984) for a good review of the macrocyst literature.
[36] See pp. 261–263 in Raper (1984) for a detailed description of this organism.

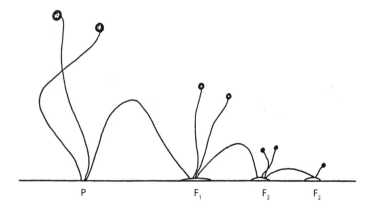

Fig. 6. A diagrammatic representation of four successive generations of stoloniferum. If the fruiting body falls over, the spores germinate and produce a new set of smaller fruiting bodies. (From Bonner et al. 1985)

aggregate and form new, smaller fruiting bodies (fig. 6). The world's record is going for six generations on an empty stomach.[37] Since there is no intake of food during this very local dispersal, the amoebae and their spores get smaller each generation, so that by the sixth generation the spore size is about the size of a nucleus of a spore of a well-fed amoeba. Yet despite its diminutive size, if bacteria are added the amoebae will engulf them and become large and thrive.

This would certainly seem to be a strategy for very local dispersal. *Stoloniferum* is found in tropical forests where the humidity on the soil or humus surface remains high, and no doubt the abundance of surface bacteria would be considerable. If this is only an effective strategy under such conditions, perhaps that would explain why it is not common, and why it is confined to the tropics.

[37] Bonner et al. (1985).

5 Behavior of Amoebae and Cell Masses

Individual Amoebae

One aspect of experimental work on cellular slime molds that has received more attention than any other is the chemotaxis of the individual amoebae. It was not known for certain until the 1940s that aggregation involved chemotaxis, although, without any evidence, the idea was put forth in the early literature. Once the mechanism was firmly established, there was the next question: What was the chemical nature of the attractant?

In the late 1960s it was discovered that the acrasin of *D. discoideum* was 3'-5'-cyclic adenosine monophosphate, or cyclic AMP.[38] As a result, the number of studies on slime mold chemotaxis has skyrocketed and continues to this very day. (Since its discovery up to today, there have been 1,190 publications on cyclic AMP in the slime molds.) The first interest was in the then recently discovered importance of

[38] Konijn et al. (1967).

cyclic AMP in hormone reactions in mammals. It was a second messenger (the first being the hormone itself) that transmitted from the hormone receptors on the surface of the cells to the internal chain of biochemical events that produced the effect of the hormone. Now lowly slime molds were using the same substance as mammals—not only as a second messenger, but also as a first messenger. At the time this caused a considerable flurry, and human physiologists suddenly wanted to know all about slime molds. However, the connection was soon realized to be remote, and as the years passed it turns out that both in mammals and in slime molds it has more than one function. It illustrates a point of considerable interest: often during the course of evolution an existing substance will be commandeered for a new role. Innovations in biochemical pathways often show the signs of opportunism; they make use of a substance that is already there serving some other purpose. As long as the initial function is not impaired, this seems to present no problem. And as we have seen, not all cellular slime molds use cyclic AMP as their acrasin; there is more than one way to produce a chemoattractant.

The first experimental evidence that aggregation in *D. discoideum* involved chemotaxis by means of the diffusion of substance, and a small molecule at that, was that of E. H. Runyon, who published a paper in 1942 in an obscure and unconventional journal.[39] He showed that if amoebae that were beginning to aggregate were put on both sides of a cellophane membrane, the opposite aggregation streams were perfectly aligned, one above the other. The key property of

[39] Runyon (1942).

such cellophane (or dialysis) membranes is that small molecules can readily diffuse through them, but larger ones, such as proteins, cannot. In other words, not only were the amoebae attracted by a chemical, but the chemical was of a small size. The case was established beyond doubt in another series of experiments published in 1947 (my doctoral thesis);[40] Runyon was absolutely correct in suggesting that his experiment was strong evidence for chemotaxis; no other explanation fit the bill.

I have written so many times the story of how cyclic AMP was discovered to be the acrasin of *D. discoideum* that I am loath to repeat it again. Let me just say it was discovered by a clever guess on the part of David Barkley, a graduate student, and Theo Konijn, a visiting colleague from Holland. Because it happened in my laboratory when I was away I have always felt some guilt: I really had nothing to do with the discovery. My only contribution was that they ordered some cyclic AMP on my grant: I paid for their crucial experiment! It was clear from the beginning that it was an effective attractant at exceedingly low concentrations. The next step was to show that the slime mold actually made and secreted cyclic AMP, which we found to be so.[41] This closed the loop: cyclic AMP was indeed was the natural acrasin.

Immediately there was a rash of work in many laboratories, and a number of interesting new things were revealed. The most important centered around the fact that the cyclic AMP was emitted in pulses, and this allowed one to follow the events in sequence. The conclusion from cleverly designed experiments was that an external puff of cyclic AMP induced

[40] Bonner (1947).
[41] Konijn et al. (1968).

a solitary aggregating amoeba to produce a cyclic AMP puff of its own; there was a relay. After the puff the amoeba would become temporarily refractory, but the one beyond would be stimulated on its own puff that would diffuse outward. This would be enough to polarize the amoeba so that it would start moving toward the center.

One matter that is known for amoebae in general is that although they look like an amorphous blob, they do have a front and hind end. This is even more evident in the elongated aggregating slime mold amoebae, where it can be plainly seen that there are different internal vacuoles and other bodies consistently placed either in front or behind the nucleus. Especially obvious is a band of contractile proteins that are evident in the posterior of the aggregating amoeba.[42]

The elongation of the amoebae during aggregation chemotaxis makes this internal polarity even more obvious. Furthermore, as I have already mentioned, the anterior and posterior ends of the elongate amoebae become sticky so that they can link up end-to-end, like a chain of elephants in the circus (see fig. 2). A gradient of the attractant clearly lines up many features at the surface and within the aggregating amoeba.[43] There have been a few reports attempting to calculate how steep a gradient is needed to orient the amoeba, calculating the difference in the concentration of cyclic AMP between the front and hind ends of the amoeba sufficient to orient. The value is very low: a 2 to 5 percent difference is enough to tell an amoeba which way to go.[44]

---

[42] Yumura et al. (1992).

[43] For a review, see Kessin (2001), pp. 131 ff.

[44] For a review, see Kessin (2001), pp. 128 ff. The first estimate of 5 percent was made by James Savage (*J. Exp. Zool. 106*: 23, 1947).

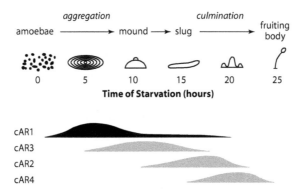

Fig. 7. A diagram showing the different times during the life cycle when the different cARs are produced. (From Parent and Devreotes, 1996. Reprinted, with permission, from the *Annual Review of Biochemistry*, volume 65 © 1996 by *Annual Reviews*, www.annualreviews.org)

The study of cyclic AMP receptors on the cell surface has been especially interesting. The cyclic AMP signals the amoeba by attaching to a special protein, and once the connection is made there is a molecular pathway inside the cell that stimulates the synthesis of more cyclic AMP. Peter Devreotes and his colleagues discovered that there are four such proteins that are related and differ slightly in their structure, and each one has a separate gene that effects its synthesis.[45] It turns out that the four are produced at different times in the life cycle (fig. 7). The first one is only concerned with aggregation; the others with later stages of development where cyclic AMP continues to play a role.

One very interesting evolutionary fall-out of this discovery was unearthed by Pauline Schaap and her colleagues.[46] They

[45] Kessin (2001), pp. 105 ff.
[46] Alvarez-Curto et al. (2005).

examined the cyclic AMP receptors in a primitive and ancient species of cellular slime mold, *D. minutum* (see the phylogenetic tree, fig. 4), and found that it only possessed one of the receptor proteins, one that was only involved in internal developmental events; their acrasin is not cyclic AMP but folic acid. This raised the reasonable idea that during the course of evolution using cyclic AMP as an acrasin was a later development and involved the recruitment of it from other, already-present pathways that involved cyclic AMP. I said earlier that there are many examples of biochemical opportunism where a substance already present was commandeered for a new, additional function. In this example of Schaap's we seem to have caught the process in the act: we have the evolutionary before and after.

There is one key element in aggregation chemotaxis that has not been mentioned. In early days, Brian Shaffer provided evidence that acrasin is very unstable and rapidly disappeared in its normal environment, but if the acrasin was collected on the other side of a cellophane membrane (that did not allow proteins to pass), it remained stable.[47] He correctly concluded that there was a protein enzyme that degraded the acrasin: an acrasinase. Also he pointed out that such an enzyme was very important for chemotaxis. If there was not some way of getting rid of the acrasin soon after it had done its work, it would become increasingly concentrated around the aggregating amoebae and soon make it impossible to maintain a detectable gradient. When it was known that the acrasin for *D. discoideum* was cyclic AMP, the acrasinase had already been well established from mammalian physiology: it was cyclic AMP phosphodiesterase.

[47] Shaffer (1956).

Since cyclic AMP is also signaling during the later stages of development, phosphodiesterase, not surprisingly, is being synthesized at later stages also. Like cyclic AMP receptors Richard Kessin and his collaborators showed that there are different phophodiesterases, each with its own gene differences.[48] One occurs during the feeding stage, another at aggregation, as I have just pointed out, and a third during the later stages of development. It would be very interesting to know if, like cyclic AMP receptor proteins, all three are not present in more ancient species, such as *D. minutum*, and that there has been an evolutionary progressive sequence there, too. This is very likely the case since *D. minutum* does not have cyclic AMP for its acrasin.

The next interesting development was the discovery of Günther Gerisch and his group that there was an additional protein that inhibited the action of the phosphodiesterase.[49] This means that the cyclic AMP is destroyed by phosphodiesterase, which then is turned off by an inhibitor, and all of these activities were occurring in the immediate environment outside of the amoebae. This can be thought of as a fine-tuning system to control the required gradients of the chemoattractant for successful aggregation. One might argue that such a complex control system is not necessary for simple aggregation, but then remember that the cellular slime molds are very ancient, and any new mechanism that reinforces or modulates a developmental process will have had plenty of time to appear, and there is no reason for natural selection to eliminate them. They are all working in the same

---

[48] For a review, see Kessin (2001), p. 158 ff.
[49] Gerish et al. (1972).

direction, that is, for effective and reliable aggregation, which leads ultimately to successful spore dispersal.

Finally, there is another important phenomenon first shown by Brian Shaffer.[50] He discovered that acrasin (these were before the days when it chemical nature was known) was placed near preaggregation amoebae, they became sticky and clumped together. The nature of these newly arising surface molecules that occur during aggregation were later intensively studied by Günther Gerisch and his colleagues.[51] These adhesive surface changes are essential for the amoebae to come together and enter their multicellular state.

FOUNDER CELLS

As we shall see, in *D. discoideum* the dominant anterior end of a slug is made up of a group of particularly active amoebae; they are the initial emitters of cyclic AMP. Brian Shaffer discovered that the situation was radically different in the two species of *Polysphondylium*.[52] There, from the beginning of aggregation, a single rounded amoeba initiated acrasin and became the dominant acrasin-producing center. He demonstrated the workings of this *founder* amoeba in a most ingenious way. Watching the beginning of aggregation through the microscope, he could see some scattered amoebae pointing towards one rounded amoeba (fig. 8). He attached a musician's tuning fork to the stage of the microscope, and as the stretched-out incoming amoebae approached the founder cell he would tweak the tuning fork. The vibrations caused the

[50] Shaffer (1957).
[51] For a review, see Kessin (2001), pp. 158 ff.
[52] Shaffer (1961).

Fig. 8. Shaffer's demonstration of a founder cell of *Polysphondylium violaceum*. Note all the other amoebae are attracted to it. (Tracing of a photograph in Shaffer 1961)

attracted amoebae to slip backward, but they soon regained their composure and started moving inward again. He found he could repeat the process a number of times with the same result. He made another interesting discovery: if he killed the founder cell with a hot needle, one of the neighboring amoebae took its place; normally a founder cell inhibited other amoebae from becoming founders. It is much like a queen bee giving off a chemical signal that controls the behavior of the sterile workers.

The two species of *Polysphondylium*—*P. violaceum* and *P. pallidum*—have both whorls and founder cells. We saw from the molecular phylogenetic tree that these two species arose separately at very different times (fig. 4). It seems puzzling that such elaborate innovations should have arisen independently a second time. However, we must remember that the tree is based on only two genes; the oddity may be resolved by further molecular analysis. Maybe whorls and founder cells are closely linked genetically and their introduction required a relatively minor genetic change.

Behavior of Multicellular Slugs

One of the most remarkable features of the cellular slime molds is the sophistication of their behavior in their multicellular stages. They are, after all, no more than a bag of amoebae encased in a thin slime sheath, yet they manage to have various behaviors that are equal to those of animals who possess muscles and nerves with ganglia, that is, simple brains. They can move; they can orient and move towards minute quantities of light (such as the dial of an old-fashioned wristwatch); they can orient in extraordinarily shallow heat gradients, a sensitivity comparable to pit vipers that can detect and strike at the warm bodies of other animals; they show fantastic sensitivity to gas gradients comparable to that of insects like mosquitoes who find us by the small amount of carbon dioxide we give off through our skin.

Obviously, one wants to know how they manage all these feats. The fact that they do it with so little specialized equipment is in itself of general biological interest.

Let me point out that many of the processes that are found in aggregation continue to play a role in the activities of multicellular stages. In the first place, the migrating slug of *D. discoideum* continues to secrete cyclic AMP where it plays an important role. Furthermore, there is a gradient of secretion: the highest output is at the anterior end, and it progressively decreases as one moves towards the rear of the slug.[53] This

---

[53] Bonner (1949).

was shown by placing a slug in the midst of amoebae that are about to aggregate and then observing their orientation towards the slug; it is clear that they point predominantly towards the anterior tip. One can also cut a slug in two and the two pieces will compete in attracting the amoebae between them (fig. 9). A small slug tip fragment will pull in a large share of the field of amoebae; the much larger posterior fragment will attract far fewer. Recently, quite by accident, I caught on time-lapse video this phenomenon on a very small slug between an agar surface with an overlay of mineral oil. The slug was moving in the oil, but it touched the agar along its lower surface that contained some aggregation competent amoebae. As can be seen in fig. 10, these amoebae streamed to the anterior end.

Since the slug tip is the high point of cyclic AMP emission, it can be considered dominant; it has all the attendant amoebae in thrall. Kenneth Raper did a classical experiment in which he grafted extra slug tips onto the sides of a large slug, and each one persuaded some of the amoebae to desert and follow the new tips (fig. 11). (I had a student who took a time-lapse film and forgot to turn it off. Fortuitously, one could see on the film a series of slugs wandering about and bumping into one another. If one slug crossed over another, its anterior end would steal the posterior amoebae of the other slug [fig. 12].) That there is a gradient of tip dominance has been clearly demonstrated; the amoebae farthest from the tip are able to defect more easily than the ones close to it.[54] Because of the tip's properties, it is sometimes called an "organizer," bearing in mind that it is quite different from Hans Spemann's organizer in the amphibian embryo.

[54] Durston (1976).

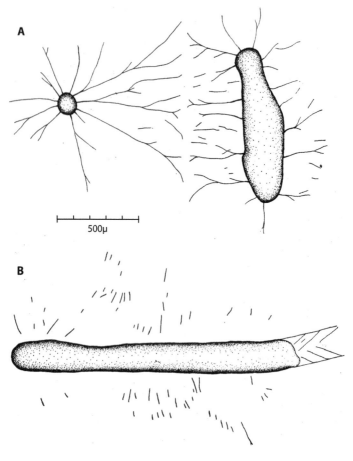

Fig. 9. Camera lucida drawings of experiments showing that the anterior end of a slug is the high point of acrasin production. The slugs are placed among receptive amoebae on a wet glass surface. (From Bonner 1949)

(A) A slug had been cut in two and the small anterior fragment attracts more amoebae and therefore produces more acrasin than all the rest of the slug.

(B) Aggregating amoebae orient towards the anterior end that is secreting more acrasin.

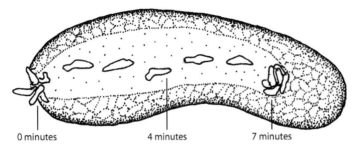

0 minutes   4 minutes   7 minutes

Fig. 10. A very small slug is in a layer of mineral oil on a glass
surface. The central clear area is where it is touching the glass. Some
aggregating amoebae on the glass-oil interface are first attracted to
the cyclic AMP emitted at the rear end (left) and then move up to
the anterior where more cAMP is secreted. They never penetrate the
slug slime sheath, but move up the gradient of cyclic AMP. This
sequence is based on a time-lapse video. The location of the forward-
moving amoebae at different times is shown. (The slug is
approximately 160 mμ long.)

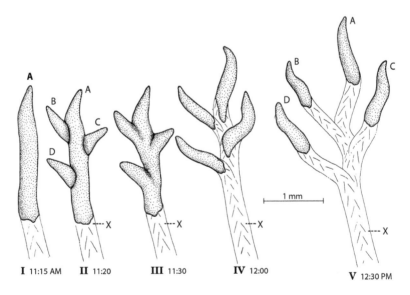

I 11:15 AM   II 11:20   III 11:30   IV 12:00   V 12:30 PM

Fig. 11. Camera lucida drawings showing that if additional tips are
grafted onto a slug, they will steal the amoebae behind them. (From
Raper 1940)

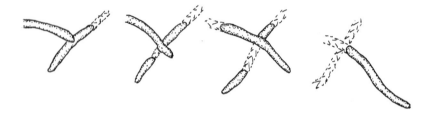

Fig. 12. Drawings from a time-lapse video showing one slug crossing another and stealing half its amoebae.

Another well-known characteristic of the tip is that it can split into two, or "twin." Also, when the tips of two slugs go roughly in the same direction and bump into one another, they will fuse and produce one larger slug (fig. 13).

An interesting thing about the acrasin emission in slugs, as in aggregation, is that it is given off in pulses. This has been shown elegantly by D. Dorman and C. J. Weijer who have recorded waves of shifts of light density starting at the tip and moving down the slug.[55] There has always been a question in my mind as to whether pulses or waves are always present either in aggregations or in slugs. The reason for my hesitation is that in time-lapse recordings of either stage, there can be periods where no pulses can be seen. This could not be clearer than in aggregation time-lapse films and also in two-dimensional slugs that I will describe presently. In the latter it could be that they are simply too difficult to detect because of their thinness. However, I do not believe there is any reason why rhythmic pulses are required; they may be more likely, but steady acrasin gradients can also do the job. However, tip dominance could also be

[55] Dorman and Weijer (2001).

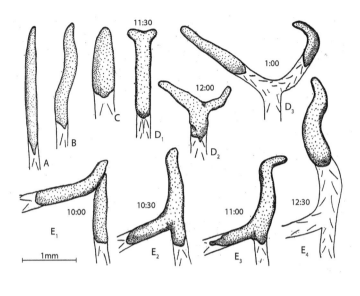

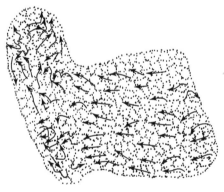

Fig. 13. *Top*: Showing twinning (D) and fusion (E) of migrating slugs. (A through C) showing normal slugs of different proportions. (From Raper 1940b)

Bottom: A minute two dimensional slug in the process of twining. Note the hyperactivity of the amoebae in both tips. (After Bonner 1998)

the result of higher frequency pulses of the cyclic AMP emission, making the dominant tip a pacemaker. In an interesting discussion, Evelyn Fox Keller points out that there is something in the human mind that is captivated by pulses and rhythms.[56]

A DESCRIPTION OF SLUG MOVEMENT

That the cyclic AMP is present in a gradient in the slug, highest at the anterior end, predisposes one to the idea that chemotaxis is as active there as it is during aggregation. Indeed, the evidence that this is so is very strong. It means that all the amoebae in the slug are oriented towards the tip, and there is no essential difference between their condition in the slugs than there is in the aggregation streams. This may be so, but there are still two major matters we want to understand: how do the slugs move as a body, and how do they orient to external signals such as light, heat, and gas gradients.

Before attacking these problems, let me digress to describe a phenomenon that I was lucky enough to fall into quite by accident. In experimental science it always amazes me how frequently an unexpected bit of luck will allow one to make a step forward. Here is what happened.

I was doing some experiments in which I was trying to get the amoebae of a slug into a very fine glass capillary by impaling a slug with the capillary and getting the amoebae to rise into the tube.[57] The point of this was to see if the trapped amoebae would show any signs of further development, and

[56] Keller (1983).
[57] Bonner et al. (1995).

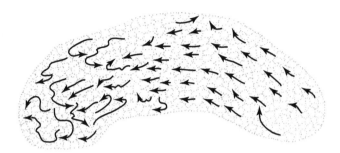

Fig. 14. A drawing of a minute two-dimensional slug, estimated to consist of about 225 amoebae (it is 175 mμ long). The figure is traced from a time lapse video. The movement of the amoebae was followed over a six-minute period, reflected by the length of the arrows. Note that the anterior cells churn about rapidly while the posterior ones are slower and go directly forward. (From Bonner 1998)

indeed they did show some before they expired. I put these capillaries with their amoebae on a glass slide and covered them with mineral oil and a glass coverslip. Oil contains far more oxygen than water, and as a result one can get almost normal development in oil. One day one of the capillaries broke right in the middle and some of the amoebae escaped and started to move along the glass-oil interface.[58] The startling thing was that it appeared to be a minute slug that crawled along with a front end and a hind end. But, even more surprising, it was a two-dimensional slug one cell layer thick; in an ordinary microscope one could see each individual amoeba. Using time-lapse photography, it was possible to follow each of them as the slug moved along (fig. 14). Seeing what was going on inside a slug or any of the multicellular

[58] Bonner (1998).

stages had always been a difficult problem; it could only be done by fixing and staining—in other words, a dead slug frozen in time—or by staining individual cells and mixing them with unstained ones and following them with time lapse. This could be viewed with the remarkable confocal microscope that takes optical slices along the slug, but the problem with all these methods is that the amoebae move up and down with the fat slug as well as forward, giving one an imperfect picture of the movements of individual amoebae. These two-dimensional slugs gave us a window into the way a slug works that had been denied to us hitherto. As normal slug movement is discussed, I will periodically intervene with what we have learned from these two-dimensional slugs.

Now for some descriptive facts about slug movement, such as their speed and the characteristics of their motion. It has been known for some time that bigger slugs move faster. A small slug will move around 0.8 millimeters per hour; a large one about 2.0 mm/hr. This can be seen in a graph from one of the best sets of measurements made by Kei Inouye and Ikuo Takeuchi (fig. 15).[59] In early days there was some discussion as to what property of the slug size was relevant: volume, surface area, or length, and these authors showed unequivocally that it was length. By causing slugs to greatly elongate in a field of static electricity, these artificially stretched-out slugs move even faster, also commensurate with their length.[60] The two-dimensional slugs are exceedingly small, yet they also show the same length-speed relation (although for their size their absolute speed is less than that of ordinary slugs, perhaps because they are moving under mineral oil).

[59] Inouye and Takeuchi (1979).
[60] Bonner (1995).

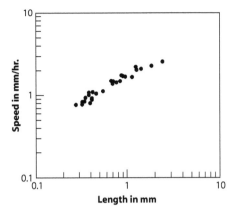

Fig. 15. The speed of migration of different size *D. discoideum* slugs. (From Inouye and Takeuchi 1979)

At this point let me introduce a parenthetical matter: one might wonder how many cells are in a slug. This was first calculated by Kenneth Raper, who estimated the volume of a slug, broke it open, and measured the volumes of some of the cells that had rounded up into spheres.[61] What he did not know was that the amoebae swell considerably as they are released from the confines of the slime sheath, and this distorted his figures. Since his results were published in the 1940s, all the many authors who followed would say that large slugs consisted of about one to two hundred thousand amoebae. The matter was reexamined using the capillaries filled with amoebae mentioned earlier, where the volume inside the cylinder could be accurately measured, and the amoebae were directly counted. It was found that Raper's estimates were about one order of magnitude off: rather than almost

---

[61] Raper (1941).

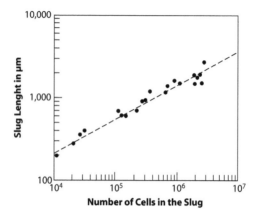

Fig. 16. The length of a slug plotted against the estimated number of amoebae it contains.

two hundred thousand amoebae, his largest slug probably was made up of one to two million amoebae (fig. 16). I published this mini-observation in the slime mold newsletter on the Web a few years ago (2001),[62] and it is interesting that all the papers that mention cell number that have been published since then still give the lower figures. On this matter, slime mold workers are conservative and like to cling to the comfortable old facts. Let me add, for comparison, the cell number for two-dimensional slugs. The smallest ever observed had about one hundred amoebae. Their number was usually less than five hundred, but occasionally one would appear that was closer to one thousand.

There has been considerable discussion over the years as to whether all the amoebae in the slug move, or whether the posterior ones push the anterior ones or the anterior ones

[62] Bonner (2001).

pull the posterior ones. It had been noted by many authors that the two regions are different, and let me say here, something that will be discussed in detail later, the larger posterior portions are destined to become the spores, and the small anterior portion is the stalk, as Kenneth Raper showed many years ago.[63] The anterior amoebae of the slug are far more active than the passive posterior amoebae, and Inouye and Takeuchi did some ingenious experiments cutting the slugs in two and noting that the isolated anterior ends moved faster than the posterior ones.[64] From these and other observations they concluded that there was greater activity in the anterior portion and that it was the locomotive for running the slug. Similar observations were made by others. Especially notable is the work of Siegert and Weijer, who showed indeed that the anterior cells were moving considerably faster than the posterior ones.[65] They also described the anterior cells as moving in a helical scroll that they imagine to be the very locomotive for forward motion.

Let us now look at all these ideas in terms of what is known from the two-dimensional slugs. In the first place it is clear, as has been generally accepted by previous work, that all the cells in a slug move, and the anterior ones move faster. In the two-dimensional slugs the anterior amoebae rush around, occasionally in a spiral that may reverse its direction, while the posterior ones move straight forward in a sedate fashion. In some cases the two regions will separate, either by touching only by a thin strand of amoebae, or separate completely, and in the latter case the posterior amoebae keep moving forward

[63] Raper (1940b).
[64] Inouye and Takeuchi (1980).
[65] Siegert and Weijer (1992).

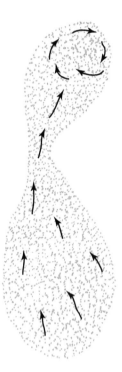

Fig. 17. A tracing from a video of a two-dimensional slug with a narrow isthmus between the anterior and posterior amoebae. The arrows indicate the direction of movement of the amoebae in the two zones.

for a time (fig. 17). They continue to be attracted to the cyclic AMP gradient. If they do not rejoin they will soon form a whirlpool of amoebae, just endlessly chasing their tails.

This observation raises an old question in my mind: Is tip dominance something more than just the high point of cyclic AMP secretion? These whirlpools are one bit of evidence that this might be so, for they suddenly are no longer aligned by the external cyclic AMP gradient which held them as long as there was a slender cell chain connecting the anterior and the posterior portions. Once that connection was broken, the ability to attract the separated group of cells lasted only a limited amount of time; then they became autonomous. The cell-to-cell connection seems necessary for dominance to be maintained. When I was a graduate student I did a reverse kind of experiment that has always puzzled me, and for that reason I never published it. I removed the center of an aggregate on the bottom of a glass dish covered with a shallow layer of salt solution, and the streams broke up. But it was curious that they did not dissemble everywhere at once; instead, a wave started near the removed center and moved progressively outward. What was particularly unexpected was that each time the wave came to a branch point of the streams there was a pause in the progressive scattering of the amoebae. It has always bothered me that some aspect of dominance involving cell contact remains a mystery. What kind of a signal passes from the tip outward, and is it in any way connected to the activities of cyclic AMP?

Individual, active anterior cells seem to churn around in all directions (although occasionally in a spiral), yet clearly at the same time they are moving forward as a group on their own; the amoebae must go in one direction more than in others. The two-dimensional slugs also tell us that all the amoebae are moving and contributing to the forward movement of the slug, and no doubt it is the cyclic AMP gradient that keeps them aligned. There is no pulling by the front end and no pushing from behind.

There is another lesson to be learned from the active cells at the tip. Because they are flat, in two dimensions, they are confined and cannot produce a scroll, but only a fleeting flat spiral. In the minute slugs in the two-dimensional experiments, occasionally one will rise up into the mineral oil and become three dimensional, and when it does it will begin a scroll at its anterior end, sometimes in one direction, and then it will reverse. This tells us that a scroll is not a requirement for forward motion; rather, it is probably an escape direction for the overactive cells.

HOW DO SLUGS MOVE?

We are now ready for the big question of how the slugs move. We have seen that all the amoebae move; we know that the slug is encased in a slime sheath; we know that there is a dominant tip and chemotaxis of the amoebae inside; we know longer slugs move faster than shorter ones. What kind of an answer can we get from all these facts?

There have been a number of theoretical studies, and they have one thing in common: they assume all the amoebae move, but to get forward they must push against something solid. One can visualize the importance of this with an analogy: a group of men lying on an ice rink, all pointed in the same direction, being able to move forward only by pushing their arms and getting leverage on their neighbors.[66] However, they would get nowhere unless the group could get some overall solid leverage, and this they could do because the men on each side will push against their wall. The slime sheath is equivalent to the wall, and in this way all the moving amoebae could manage a collective forward movement. This model could be modified by assuming that all the individual internal amoebae find

[66] Odell and Bonner (1986).

something firm to push against. There are a number of possibilities. For instance, the slime sheath is secreted by all the amoebae, including the internal ones and not just those at the surface: this could be a firm substratum to push against.

Recently I began looking at another unrelated—and more ancient—species of slime mold, *Dictyostelium polycephalum*, that also has a stalkless migration slug. It is different from *D. discoideum* in many interesting respects. The slug of *D. polycephalum* is much thinner and longer; it is rather like a strand of spaghetti (fig.18). These slugs will migrate for great distances in soil, and unlike any other species they can move through a thick layer of overlying soil, and can even migrate short distances through agar, a unique accomplishment.[67] Unlike other large *Dictyostelium* species, this one does not have a special anterior zone: the amoebae appear to be uniform along its slug axis and there is no anterior zone of hyperactivity.

It also follows the rule that longer slugs move faster, although its rate of movement is considerably slower than *D. discoideum*. It is possible to cut a slug in two and measure the speed of each half, even when the cut is made so that the two parts are of a very different size. The rate of movement of whole slugs and of all these fragments is measured, and they clearly follow the length-speed rule, although the data are scattered (fig. 19). The important point is that here we have the same relationship as was found in *D. discoideum*, but there is no highly active anterior zone. That special anterior region is not needed for slug locomotion.

From the peculiarities of the migration of this species and the migration in two dimensions of *D. discoideum* we have

[67] Bonner (2006).

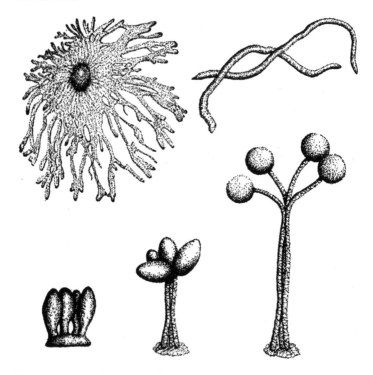

Fig. 18. *Dictyostelium polycephalum*. *Upper left*: an aggregation. *Upper right*: migrating slugs. *Bottom*: fruiting, in which one slug breaks up into a number of small fruiting bodies. (Drawings by R. Gillmor based on photographs from Raper 1956)

some new ways of looking into how slugs move. All the amoebae move; the active anterior zone is not a necessity for that locomotion, and that includes the helical scroll; any portion of a slug can obey the length-speed rule. It is this latter point that seems to me so intriguing. I have asked various mathematician and physicist friends about it and they have not been too helpful. This led me to an analogy that I find useful (although it does not ignite enthusiasm in those

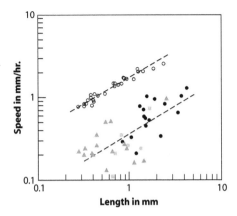

Fig. 19. Log-log graph showing the relation between slug length and speed. *Top*: Data from Inouye and Takeuchi (1979) for *D. discoideum*. *Bottom*: Data for *D. polycephalum*. Solid circles are for intact slugs; triangles for severed tips; and squares for severed posterior ends. *D. discoideum slugs* move roughly five times faster than *D. polycephalum*. (From Bonner 2006)

friends). Thomas McMahon compared the speed of racing shells with different numbers of oarsmen (1,2,4,8) and showed that the more oarsmen, the faster the boat.[68] If one thinks of each amoeba as an oarsman, then the more amoebae, one in front of the other, the greater the speed. This may be right but the difficulty of arguing it in detail is daunting to say the least. All the forces involved, such as the power of each amoeba, drag, and other matters of fluid dynamics, are extremely complex and hard to measure in something so small as an amoeba: rowing shells are much easier!

Another set of experiments has shed some further light on the matter. In an unpublished study done some years ago,

[68] McMahon (1971).

Michael Macko measured the speed of movement of haploid and diploid strains of *D. discoideum*, for both separate amoebae and slugs.[69] Diploid strains have twice the number of chromosomes, and their cell size is roughly twice that of the haploid strain, which has only one set (which is the normal condition of slime molds in nature.) He found that the speed of movement of separate larger amoebae is considerably faster than that of the smaller ones, yet if we compare the speed of slugs of the same size of the two strains, we find they move at the same speed. The diploid slugs have approximately half the number of cells of the haploid ones, but their amoebae move much faster. This would help to explain why haploid and diploid slugs of the same size move at the same speed. The implication is that the critical element is the total mass of cytoplasm regardless of the size of the units in which it is packaged. It would be the total amount of contractile protein that would correlate with speed—the longer the column of that contracting motor, the faster it moves. This lesson from *D. discoideum* presumably applies directly to what is found in *D. polycephalum*, where the speed is proportional to the length of the column of amoebae.

Orientation

With the discussion of slug movement and how it might be achieved behind us, we now come to the intriguing matter of how a slug orients in its environment—how it manages to go towards some things and away from others. In nature slugs must manage to migrate from underneath the soil to

[69] Macko (1971).

the surface so they can be dispersed by small passing animals. As I have already mentioned, and now we shall see in detail, they orient to light, to heat, and to external gases with amazing sensitivity.

The first question is, how does a slug turn to go in a new direction? For years I assumed this was not a matter that needed to be seriously discussed: the answer was so obvious. And now all the evidence shows that I have been dead wrong all this time. I assumed, since all the amoebae in the slug were moving, the amoebae on one side of the slug only needed to move faster than those on the other to make a turn. This seemed so totally obvious that I could not even imagine there was an alternative. However, it has been pointed out by Paul Fisher and collaborators that if one shifted the dominant cyclic AMP emitting tip of the slug to one side, the amoebae behind it would follow in the new direction.[70] This is supported by what we see in the two-dimensional slugs: one cannot detect any difference in speed of the amoebae on the two sides of a center that is shifting its position.

AMMONIA

Our first step is to understand what makes a slug tip turn. If it is the result of the shifting of the location of the dominant cyclic AMP-producing center, then what causes the shift? One answer, and probably the crucial one, is ammonia gas.

Some studies have shown that the production of cyclic AMP is inhibited by ammonia.[71] Initially these were rather

---

[70] Fisher et al. (1984).
[71] Schindler and Sussman (1977); Thadani et al. (1977).

gross biochemical experiments in which it was simply demonstrated that in the presence of ammonia less cyclic AMP is produced. One of the effects of this inhibition is to prevent fruiting: in the presence of ammonia gas, migrating slugs will not form fruiting bodies but will continue to migrate endlessly. This leads to the idea that the cyclic AMP-dominant center in the tip, which is made up of a bunch of very active amoebae, will move away from the higher to the lower concentrations of ammonia.

Ammonia is constantly given off by cells, for it is the normal by-product of the breakdown of proteins and other nitrogenous substances that goes on constantly in the biochemical cycling within cells. Being such a small molecule, it diffuses extremely rapidly. There is not only a transient gradient that is highest at the surface of the slug, but in the soil there are all sorts of other organisms giving off ammonia—bacteria, algae, worms, and innumerable other beasties. So there will be many ammonia gradients in the soil surrounding a slug, and this is one of the factors that will guide its migration movement towards the surface. Let me now give the evidence that ammonia is a key player in slug orientation.

The first inkling came from evidence that rising fruiting bodies were affected, in fact repelled, by a gas. It had been noted that when two fruiting bodies rose into the air close together, they leaned away from one another: why was this so?[72] One hypothesis was that they gave off a repellent gas, but at first this only seemed to be a wild and unlikely bit of speculation. However, it was soon supported by some experiments. The crucial one was that if a small heap of activated

[72] Bonner and Dodd (1962a).

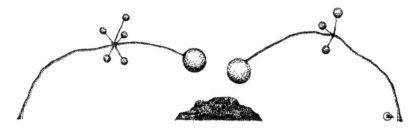

Fig. 20. Two fruiting bodies of *Polysphondylium pallidum* leaning into a heap of activated charcoal. (From Bonner and Dodd 1962a. Drawing by R. Gillmor)

charcoal (which adsorbs gases, including ammonia, with great avidity) was placed beside a rising fruiting body, the whole fruiting body, as it rose, turned directly into the charcoal (fig. 20). Because the charcoal removed the repellent gas, there was less of it on the charcoal side than on the opposite side, and therefore the rising fruiting body plunged into the charcoal. This means that two rising fruiting bodies moved away from each other because the space between them had a high concentration of the gas.

Then came the question of what was the gas, and it turned out to be ammonia.[73] The must convincing demonstration of this was made by T. Kosugi and K. Inouye, who made a very fine glass pipette that gave out minute amounts of ammonia.[74] It was arranged on a micromanipulator so that the tiny glass tip could by positioned on either side of the tip of a migrating slug. If the gas-emitting tip was on the right, the slug tip would turn to the left, and vice versa.

[73] Bonner et al. (1986).
[74] Kosugi and Inouye (1989).

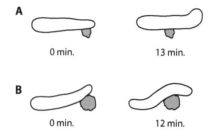

Fig. 21. Effects of a protease and a protease inhibitor on the orientation of a migrating slug.
(A) A polyacrylic bead saturated with a protease is placed near the slug and after a delayed response the tip moves away from where the protease was originally placed.
(B) A mixture of protease inhibitors was similarly placed on a slug which curls around the bead. (Drawings from video frames, in Bonner 1993)

That the slugs themselves could produce the right amount of ammonia to do this was shown in another way.[75] Minute bits of acrylic gel were soaked with a proteolytic enzyme that would break down proteins and liberate ammonia. The gel was lodged on the side of a slug migrating on agar to break down some of its surface proteins, and after ten minutes or so the slug suddenly took a sharp turn away from the side with the gel (fig. 21A). A reciprocal experiment consisted of putting a protease inhibitor in the acrylic gel, and then the tip turned towards the gel side (fig. 21B). There had to be less ammonia on the gel side because the normal breakdown of proteins on the slug surface had been inhibited. All this adds up to pretty strong evidence that slugs are very sensitive to gradients of ammonia gas and orient to those gradients.

[75] Bonner (1993).

OXYGEN

In a different line of attack, John Sternfeld and Charles David discovered that slime mold cell masses were also sensitive to oxygen gradients.[76] They put cells that were close to aggregation in depressions on an agar surface and covered them with a glass coverslip; they then put the whole preparation in a chamber filled with 100 percent oxygen. Despite the cells' confinement, they developed a front and a hind end, and their front ends invariably pointed towards the edge that was closest to the oxygen atmosphere; they were orienting in an oxygen gradient. Unlike ammonia that repels, oxygen was doing the opposite: it was attracting the amoebae, and indeed was behaving like cyclic AMP.

The significance of these results was slow in being fully appreciated. Later there was an astonishing experiment by M. Yamamoto that illustrated how important oxygen gradients are.[77] He used a technique that is very clever, and so far as I know it remains in the unique province of Japanese workers. Hot agar is poured around a thin glass needle, and, after cooling, the needle is extracted, leaving tunnels about the diameter of a slug. Next, and this is the part that I always thought requires special talent, a slug is persuaded to enter and migrate into the tunnel. Yamamoto's slugs were stained with the vital dye Neutral Red that fortuitously stains the anterior cells dark red. The tunnels he used were short with a dead end, and the slug migrated right into the cul de sac. All further forward movement was blocked, but a truly amazing sequel occurred. The slug essentially turned around and came

[76] Sternfeld and David (1981).
[77] Yamamoto (1977).

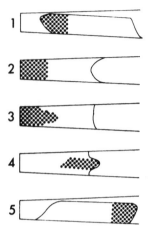

Fig. 22. A diagram of how slugs, with their anterior cells stained with a vital dye, reverse their polarity and direction of migration when crawling into an agar tunnel. (From Yamamoto 1977)

back out of the tunnel. However, it did not do this by the whole slug making a U-turn; instead, all the anterior darkly stained cells percolated through the previously posterior, relatively colorless cells, and a new red front end emerged from the tunnel opening (fig. 22). The dominant cyclic AMP tip was reestablished at what was the previous posterior end; apparently the oxygen gradient overrode the old cyclic AMP gradient, and a new one was established. Each individual amoeba made a U-turn and now obeyed its new master in the opposite direction. So, oxygen gradients can orient slugs, and they do so by placing the position of the dominant center at the high end of the oxygen gradient, which in turn dictates the direction of the cyclic AMP gradient. It should be noted that since each individual amoeba acts on its own, rather than the whole slug, a slug is not quite as unified as a multicellular metazoan; by comparison, the amoebae in a slug show a unique degree of democratic independence.

As with everything in this book, the picture I have painted of the role of gases in the orientation of slugs is very much my own idiosyncratic view. Many of my slime mold colleagues would no doubt have written it up differently.[78]

Let me add parenthetically that there is yet another gas that plays a role in the life cycle of cellular slime molds. Ethylene is not responsible for orientation, but for the induction of macrocysts, the sexual phase, and the resting body that appears to be a safe haven in flooding.[79] Ethylene is known to be a common gas emitted by higher plants. It not only stimulates plant growth but it is responsible for fruit ripening: it is the substance that hastens bananas to turn yellow. It is interesting that these small molecules that are volatile substances should play such a prominent role in the social soil amoebae. Perhaps it is related to the fact that gases can penetrate soil more rapidly and effectively than larger molecules. Even the various acrasins are all relatively small molecules, although not small enough to be volatile. Each signaling substance would appear to be the right size for its respective job.

LIGHT

One of the most striking phenomena is the ability of slugs to orient towards light. When one arrives in the laboratory in

---

[78] For instance, there is a rival story here. Paul Fisher (1991) argues that there is a substance (slug turning factor) that is produced by the amoebae and is responsible for inhibiting and moving the dominant center at the tip of the slug. I must confess I am not a fan of this factor. As I make clear in the text, I push for ammonia.

[79] Amagai (1984).

the morning, the slugs in every culture dish will all be pointing towards the window. They are strikingly phototactic. The
amount of light needed for this orientation is very small;
they are extremely sensitive to a spot of light of very low intensity. If a dish is placed between two very dim light bulbs
eight feet apart in a dark room, and if the slugs are exactly in
the midpoint between the bulbs, they will wander about in a
confused fashion; but if they are one or two feet closer to
one of the bulbs, that slight difference in light intensity is
enough to have them go towards the nearer light.[80]

The question how the slug senses the light and responds
by turning immediately arises. It is known that there is a specific pigment that absorbs the light, but we do not know the
biochemical pathway that leads to turning. Instead, we have
somewhat more general information of the steps involved.

In the early 1900s J. Buder did some fascinating work on
the orientation to light of fungi that had fruiting bodies that
were a single filament that leaned into the light coming from
one side.[81] He showed that this was a "lens effect"; the cylindrical filament that was transparent inside served as a lens
and focused the light on the farther side of the tube. This
concentration of light on the back side stimulated its increased elongation, its growth, and the filament pointed into
the light. To prove his point he did two experiments. In one
he made it so the inside of the filament was not transparent,
but opaque, with the result that the front had more light
than the back and the filament grew away from the light. In
another he put the rising fruiting body in mineral oil, where
it thrived, but because of the difference in the refractive

---

[80] Bonner et al. (1950).
[81] Buder (1920).

index of oil as compared to air, the lens becomes a diverging rather than a converging lens, making the front more illuminated than the back, and again it moved away from the light. In both cases the brighter side grew faster (fig. 23).

Slime mold slugs do not grow; they move. Yet it is possible to do the very same experiments and they show the same result. They lose their translucency if the cells contain a vital dye, such as Neutral Red, and as a result they turn away from the light. They also turn away from the light if they are migrating in mineral oil. In some definitive experiments, David Francis directed a minute focused beam of light on one or the other side of a slug tip, and the slug always moved away from the illuminated side (fig. 24).[82]

There is some doubt and confusion as to exactly what the light is doing. It is known that light can increase the amount of ammonia in a slug, but there is also some evidence that it stimulates the production of cyclic AMP, which would seem to be contradictory phenomena.[83] A local increase in ammonia would explain the shifting of the dominant center to one side, but how does this square with the stimulating effect of cyclic AMP? Clearly this is something for future research.

It is important to mention that not all species of cellular slime molds are capable of orienting to light. In fact, Pauline Schaap and her and colleagues report that among the 59 species studied, 37 species strongly orient, 8 weakly orient, and 14 don't orient at all.[84] Since it is assumed that phototaxis serves slime molds in the soil as a way to reach the surface in order to fruit—an essential part of their life history—there

[82] Francis (1964).
[83] Bonner et al. (1988); Miura and Siegert (1988).
[84] Schaap et al. (2006).

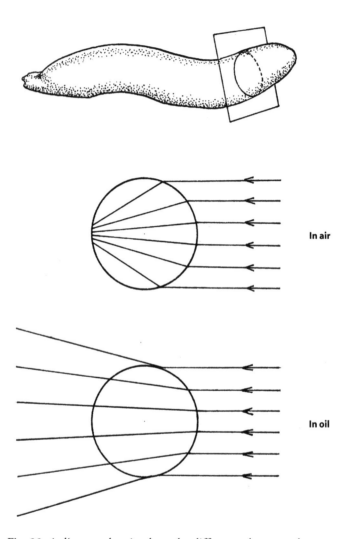

In air

In oil

Fig. 23. A diagram showing how the differences between the refractive index of water and oil affect the direction of the light rays across a translucent slime mold slug.

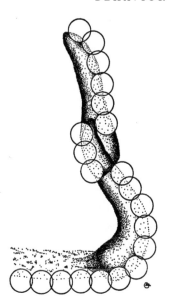

Fig. 24. The effect of a spot of light on the direction of movement of a migrating slug of *D. discoideum*. *Left*: Showing where the spot was put on the slug. *Right*: Showing the effect of the spot put successively on one side and then on the other. The slug always turns away from the illuminated side. (Redrawn from Francis 1964, by R. Gillmor)

must be other strategies that bring slime molds to the top of the soil. Indeed, they do exist: gas gradients are always present in the soil, as are temperature gradients.

HEAT

The person who discovered that slime mold slugs orient in temperature gradients was Charles Neeley, an undergraduate doing his senior thesis research.[85] With the simplest and

[85] Bonner et al. (1950).

crudest equipment he was able to show that slugs would go toward the warmer side even when the difference in temperature hitting it from the two sides was minute. (A difference of 0.0005°C between the sides of a small slug was a sufficient clue for it to move in the warmer direction.) This was met with considerable skepticism that taught me an early lesson. If a result is greeted with disbelief, or even scorn by the outside world, there is a good chance that it is not only true, but important. Fortunately, a few years later Kenneth Poff repeated the experiment using beautifully designed and sophisticated equipment, and the extreme sensitivity to heat gradients was confirmed.[86]

Then Poff and his student B. D. Whitaker followed up this result with a really important discovery.[87] They found that slugs only did this if the experiment was run at an overall temperature greater than the temperature at which they were grown; if they were tested in a gradient at cooler temperatures than the growth temperature, the slugs migrated to the cooler side of the gradient; they were negatively (rather than positively) thermotactic. This astonishing result led them to the following conclusion as to what might be happening in the soil: In the daytime the overall temperature was on the warm side with the gradient warmest at the surface, and therefore the slugs would migrate upwards. At night the overall soil temperature would be cool and the temperature gradient reversed: the soil would still have some of the residual warmth from the day, while the night air would cool the surface. This means that in the dark, because of the reversal of their

---

[86] Poff and Skokut (1977).
[87] Whitaker and Poff (1980).

chemotactic sensitivities, the slugs would continue to migrate to the soil surface. They are programmed to migrate upwards night and day.

NEUTRALIZING SLUG ORIENTATION

To me one of the most interesting observations is that all the orientations just discussed, to light, heat, and gas gradients, can be eliminated in an atmosphere of ammonia. In a chamber containing the appropriate concentration of ammonia gas, the slugs fail to respond to light or to temperature gradients; their ability to sense these stimuli vanishes.[88] Perhaps the reason is that the external ammonia drowns out the effect of the local ammonia gradients within the slug that push the dominant cyclic AMP-emitting center at the tip to one side or the other. What is more, the ammonia atmosphere allows gravity to take over.[89] If a slug is on an agar surface in an inverted culture dish (that is, it is crawling along the ceiling) in a chamber with ammonia gas, then any resistance to gravity is lost and the slug droops downward, even forming hanging fruiting bodies (fig. 25).

ORIENTATION IN THE SOIL

One must go back to the question of what these orientations are doing for the slime molds in their natural habitat, the soil. There is clear evidence that they are all part of a plot to get the fruiting bodies to the surface so that they can be dispersed

---

[88] Bonner et al. (1988).

[89] It is of interest that Häder and Hansel (1991) have shown that slugs are affected by gravity.

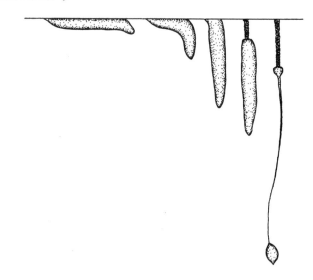

Fig. 25. In an atmosphere of ammonia gas, slugs and fruiting bodies lose their ability to orient and gravity takes over. (From Bonner et al. 1988)

by passing invertebrates. Intensive selection pressure has been favoring all mechanisms that facilitate dispersal. In some ways it is surprising that there are so many different strategies, responses to so many different cues. As we have seen, slime mold slugs respond to light, temperature, and gas gradients; all of them seem to be designed for the same purpose. Does this not seem to be a bit excessive?

Perhaps not when one considers it from an evolutionary point of view. In the first place, these slime molds have been in existence for many millions of years, more than enough for many features that are under compelling selection to evolve. As I have emphasized repeatedly, there is strong selection pressure for dispersal. This is the feature that allows successful reproduction of successive generations, the essential fruit of natural selection. For argument's sake, let us say that the first system to evolve was light to steer the slugs to the surface. If by chance another system evolves, such as a response to a gas gradient that guides the slug in the same direction both in the night and in the day, it will certainly not be selected against. If getting to the surface is so vital for dispersal, then any new invention that arises and does the same thing will persist. It is an insurance against failure of the light response, a back-up. And the same argument could be made for gas gradients.

There is the interesting question of which came first: light, heat, or gas. We do know that there are a number of species that do not respond to light, but we do not know if that is whether they never had the gift, or they lost it. Also, we do not know what species have temperature or gas gradient sensitivities; no one has looked. Most of the work has been done on very few species.

One also must wonder at the complexity of some of these orientations. In the case of gas orientation the slugs are attracted to the surface because there are two gradients. Oxygen is highest at the surface and this attracts slugs upwards; ammonia is highest in the interior of the soil, where it is being emitted by myriads of bacteria and other minute organisms, and since it repels slugs they escape to the surface. There is good evidence that both of these gradients do actively participate in getting the slugs to the surface.[90] An even more sophisticated example is that of orientation to temperature gradients. The fact that there can be a reversal at night and that negative thermotaxis supplants the positive

---

[90] Bonner and Lamont (2005).

thermotaxis of daytime seems an extraordinarily complex adaptation just to be sure to get to the surface of the soil. No doubt all those millions of years of evolution help, but it still stretches the imagination that the mechanisms can be so intricate and delicate in a mere bag of amoebae.

6 MORPHOGENESIS

The Variety of Shapes

The variety of shapes of cellular slime molds has been admirably summarized by Pauline Schaap and Sandie Baldauf and their colleagues: as was discussed earlier, they did a molecular phylogeny of seventy-five species and separated them into four groups. The largest and most recent species were members of group 4 (fig. 4).[91] They consist mostly of simple fruiting bodies with a single stalk capped by one terminal spore mass. The model slime mold *D. discoideum* is a member of this group.

Another member of the group is an interesting exception, for it has multiple sori. *D. rosarium* is a fairly large species, and as it moves upward it buds off groups of amoebae at the rear end of the rising cell mass. But unlike *Polysphondylium*, these fairly regular groups do not re-form to produce whorls, but all the amoebae in them turn into spores (fig. 26). They are indeed like beads on a rosary.

[91] Schaap et al. (2006).

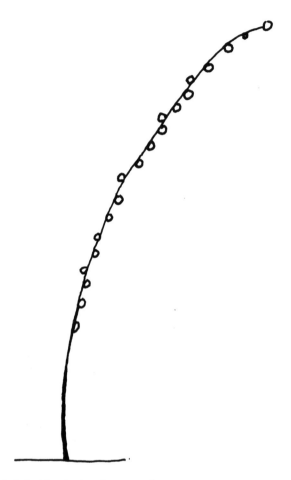

Fig. 26. A fruiting body of *Dictoystelium rosarium*. These can be over 1 cm in length, but when emerging over soil they tend to orient themselves horizontally, presumably because they are attracted to the soil. (See Bonner and Lamont 2005)

The other exception is *Polysphondylium violaceum,* whose position on the phylogenetic tree lies at the edge of group 4. It also is a large species and it has beautifully regular whorls. Like *D. rosarium,* groups of amoebae are pinched off posteriorly as the fruiting body rises, but instead of all the amoebae in the group becoming spores, they regroup and form miniature secondary fruiting bodies that develop outwards from the main stalk like spokes on a wagon wheel, each with a minute terminal sorus. As mentioned before, *P. pallidum* has an almost identical morphology, but it is found in the more ancient group 2.

The mechanism of the whorl formation has been examined in some detail by Edward Cox and other members of his laboratory.[92] By combining mathematical modeling with experimental observations, they were able to gain an understanding of how the beautiful architecture is achieved. They show that the pattern can be understood in terms of reaction-diffusion phenomena, and that the formation of the whorls is controlled by the shape and size of the surface of the mass of cells that gives rise to the whorls. During the course of this work they made two additional discoveries that seem to me of particular interest.

One has to do with comparing fruiting bodies of the normal haploid strain (with one set of chromosomes) with a diploid strain (with a double set of chromosomes, whose cells are twice the normal size). Their fruiting bodies appeared identical: their total size was the same as were the distances between the whorls. This means the number of cells in the stalk between the whorls in diploid fruiting bodies had half

[92] See Cox et al. (1988); Spiegel and Cox (1980); McNally et al. (1987).

the number of stalk cells of those in the equidistant haploid fruiting bodies. In other words, it is not the cell number that is important, but the amount of protoplasm (cytoplasm and nuclei) is key to marking off the distances. Note that we came to this same conclusion in discussing the rate of movement of haploid and diploid migrating slugs of the same size that moved at the same speed regardless of their component cell size. It should also be pointed out that it fits in well with the beautiful experiments of Gerhard Fankhauser on newts with cells of different sizes (and chromosome numbers), where again the composition of organs of the same size can be constructed with cells of widely different sizes.[93]

The other observation that Cox and his group made is that the rate of the rising fruiting body into the air in *P. pallidum* remains essentially the same until just before it stops.[94] The situation is quite different in *D. discoideum*, where, as it culminates, the speed progressively decreases.[95] As we shall see, this difference is no doubt due to the early differentiation of the amoebae in *D. discoideum*.

Size

There are many species in groups 1, 2, and 3 that do not have a solitary fruiting body but instead break up into smaller structures in various ways. They do this by producing multiple tips, often in the form of branches. In the branched species they either form regular whorls, as in *Polysphondylium*, or the

[93] Fankhauser (1955).
[94] Cox et al. (1988).
[95] Bonner and Eldredge (1945).

branches are irregular and haphazard. This appears to be akin to the twining we saw in slugs, where a dominant tip splits in two, and the splitting occurs in the rising stalk tip (see fig. 13). Occasionally a small stalked slug will climb up the side of a larger stalk and give the appearance of a branch, but in reality it is just another individual.

Multiple tips are also formed in other ways. Often the aggregation streams will break up to form numerous new dominant tips, each one of which will produce a small fruiting body. In other cases the central mound of the aggregate will break up into a group of smaller dominant tips, and again each one produces a separate fruiting body.

D. polycephalum is particularly interesting in this regard for it produces a new set of dominant tips twice in its development. At the end of a rather loose aggregation, instead of one central tip a cluster of tips arises. Each one develops into a long thin slug, and, after an extended migration, its anterior end will again break up into multiple tips (usually somewhere between three and seven) so that a number of fruiting bodies are formed that are very close together (see fig. 18). As they rise up into the air, they stick to one another for most of the upward journey, although near the end they separate and flare out, each one with a terminal sorus. They look rather like a bunch of minute flowers in a thin vase.

CONTROL OF SIZE

The above descriptions raise the question of how size is controlled. All the experimental observations on the mechanisms of size control have been done with *D. discoideum*, but the results are probably relevant to other species as well.

From the very beginning it was obvious that there must be a control mechanism that determines the size of a species' fruiting body. As Kenneth Raper showed in his monograph of the dictyostelids, one of the diagnostic features of a species is its height, which is a reflection of its size.[96] Even though there must be a control mechanism, for any one species the size range is considerable.

The reason for this is that for slime molds size is determined in quite a different way from organisms that grow from an egg and enlarge until they stop, as indeed we do as human beings. Because in slime molds growth occurs first by the eating and the multiplication of separate amoebae, initially the size for the multicellular stages is determined by how many amoebae enter an aggregate. In the laboratory the size of an aggregate can further be controlled by the density of the starved amoebae on a culture dish. If they are dense, one can have aggregates of a million or more cells; if they are sparse, they can consist of just a few cells. The smallest are the *Protostelids*, organisms that are distantly related to the multicellular slime molds. They consist of a single amoeba that secretes a minuscule stalk to raise a single spore into the air (fig. 27).

For some time it has been known that occasionally mutants of *D. discoideum* will produce very small fruiting bodies.[97] These "petite" mutants arise by the breaking up of streams, as mentioned earlier. The lesson for us here is that there must be a genetic control of slime mold size, and indeed the same conclusion must follow from the differences in species sizes just

[96] Raper (1984).
[97] Hohl and Raper (1964); Kopachik (1982).

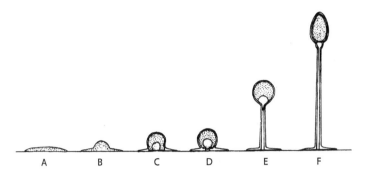

Fig. 27. A diagrammatic representation of fruiting in a species of *Protostelid*. A single amoeba secretes a stalk and becomes a spore as it rises up into the air. (From Olive 1975)

mentioned. However, one still wants to know the immediate control mechanism; what do the size genes turn on and off?

The first to examine this question was Will Kopachik, who discovered that two chemical messages were involved within cell masses of *D. discoideum*. One was that known from Raper's earlier work: the ability of a dominant tip to suppress secondary tips. Now it turned out there was a responding system that also involved the diffusion of a factor that inhibited the effect of the dominating factor; it was a two-way chemical conversation. He was able to show that the difference between the "petite" mutant and the normal strains was that the smaller strain had a greater resistance to the influence of the dominant tip; it produced an inhibitor substance that counteracted the dominating substance. No doubt the dominating mechanism involves the cyclic AMP relay-gradient system.

This matter was taken up more recently by Richard Gomer and his associates. They showed that the process was a form of quorum sensing, a phenomenon of current interest in microbes, especially among bacteria.[98] The amoebae produce a substance during aggregation that becomes increasingly concentrated the greater the number of amoebae, and when it reaches a certain level it causes the breakup of the aggregation streams to produce smaller fruiting bodies. They have gone into the molecular aspects of this phenomenon and find that the substance is made up of polypeptides whose activities are entwined with the cyclic AMP machinery.

It is not clear how this quorum sensing factor fits in with Kopachik's repressor substance. Both work in conjunction with the cyclic AMP system of dominance, but they are obviously different. There is no reason why both could not be involved. Those working on the molecular genetical analysis of the matter will no doubt eventually clarify the detailed mechanism, but if one looks at the overall problem one can only be impressed by the enormous complexity of the steps to guide such a seemingly simple matter as the control of size in the cellular slime molds.

---

[98] Gomer and his collaborators have written quite extensively on this subject. For a brief review, see Gomer (1999).

7 DIFFERENTIATION

For the cellular slime molds, differentiation involves the division of labor between stalk cells and spores. It is true that some species have at the base of their stalk special structures such as a basal disc or a crampon anchor that are made up of vacuolated cells similar to stalk cells; they represent another way in which the labor is divided. However, I will stick to the main stalk-spore division. It clearly is the matter of central interest, and a vast literature on the subject has accumulated. Not only are there fascinating matters of descriptive and general biological interest, but this is one area where molecular biology has been a crucial tool in the general search for mechanisms.

But before beginning on our main quest I should say something about an ancient genus (in group 2, fig. 4) that has only one cell type. *Acytostelium* is unique in this respect: all the amoebae in the rising fruiting body first secrete a thin, acellular stalk and then all turn into spores (fig. 28). This has seemed to me a bit of a paradox because the argument has always been that all other stalks are made up of stiffened

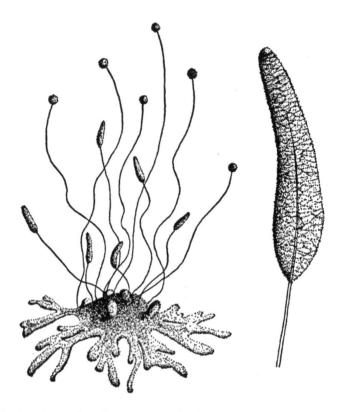

Fig. 28. *Acytostelium leptosomum*. *Left*: The simultaneous occurrence of aggregation and culmination. *Right*: An enlarged slug showing the acellular stalk and the cell orientation. (Drawings by R. Gillmor, based primarily on photographs by Raper and Quinlan 1958)

vacuolated cells that provide strength in the form of cross-struts to lift the sorus up into the air. But in *Acytostelium* this is managed with a solid cylinder of cellulose. It is true that all its species are small, and obviously its fruiting bodies manage dispersal very well and have done so for millions of years.

Since almost all the experimental work on differentiation has been done on *D. discoideum*, let me briefly describe how it arises in the life cycle. At the end of aggregation the slug is uniform in its appearance.[99] There are no special zones of different composition; in this regard it is like the slug of *D. polycephalum*. We have seen that the cyclic AMP secretion is fairly uniform down its length except for the tip, where it gives off more (see fig. 9). After anywhere from less than an hour to five hours later, it will suddenly become radically different. There arises a sharp, distinct division line between an anterior and a posterior zone; very roughly the posterior zone is about three times larger than the anterior. As migration continues, these two zones persist, and the amoebae in the anterior one become stalk cells, while those on the posterior zone will become spores. For this reason they are called "prestalk" and "prespore" amoebae.

At first these zones could be seen only in fixed and stained preparations, but fortunately it was found that if the amoebae before aggregation were treated with a nontoxic or vital dye, such as Neutral Red or Nile Blue Sulphate, the anterior zone showed much more intense staining than the posterior zone, which can even be seen clearly in two-dimensional slugs (fig. 29). Right after aggregation, slugs are uniformly

[99] Bonner et al. (1990).

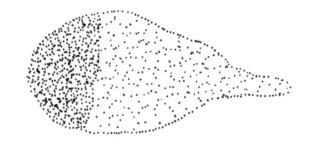

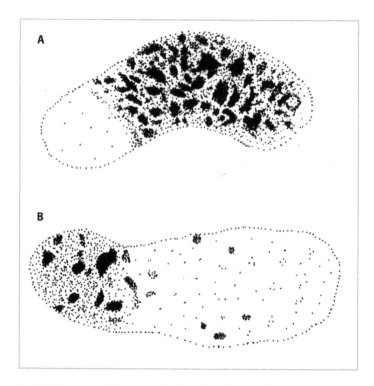

Fig. 29. Tracing of photographs showing the prestalk-prespore division in two dimensional slugs. *Above*: Neutral Red clearly stains the anterior prestalk cells. *Below*, (A): A slug stained with green fluorescent protein attached to the promoter of a prespore protein; (B): the dye on a promoter attached to a prestalk protein. (Note that a few anterior like cells appear in the posterior prespore region.) (From Bonner et al. 1999)

stained, and at some time within the first five hours of normal migration they become two-toned.[100]

If we look at other species of slime molds to see to what extent they show this differential staining effect with vital dyes, we find that it is present in the other big species, for example in *D. mucoroides* and similar species, but totally absent from *Polysphondylium*. In the case of the latter, one appealing argument is that since the cell groups break up into whorls with new stalks, they must not advance too far in their development since they would have to backtrack and start anew with each secondary branch. The same argument applies to *D. polycephalum*, whose slugs also show uniform staining with a vital dye; there is no dark tip. And at the end of their migration, the anterior end breaks up into a number of new tips that form a tight cluster of small fruiting bodies. Again, by delaying the prestalk and prespore partial differentiation they can go continuously forward in their path to final differentiation.

There may be something to this, but there is another largely untested matter. All the large species in group 4 (fig. 4) that have been examined show the two zones, and although there has been no systematic testing of species in the other three older groups, from the little we know it is conceivable that no species from those groups show the two-toned staining. So one of the evolutionary advances to be found in group 4 is not only a size increase, but possibly also the invention of prestalk and prespore cells, a sort of halfway point in the differentiation process. As we can see in figure 4, *Polysphondylium* lies at the beginning of group 4. So there are two

[100] Ibid.

hypotheses: one is that the prestalk-prespore development had not yet have been invented; the other, as I have argued above, it is mechanically awkward and inefficient to have the preliminary presumptive zones arise early when later in development they would have to reverse because multiple tips are yet to be formed, as in the production of whorls.

The function of these zones arising in early development remains puzzling since the species that do not have them seem to manage perfectly well in producing both spores and stalk cells, and do so in roughly the same proportions as those with the zones. This would seem to argue that this halfway step in differentiation is no great evolutionary step forward, but something related to the timing of the differentiation process. Perhaps these early steps in differentiation do occur in *D. polycephalum* and *Polysphondylium*, but it is not a process spread out in time; it takes place very rapidly just at the last moment when it is needed. This means that in *D. discoideum* and similar species, the steps in their differentiation are spread out over a period of time where we can see them and experiment on them. In other words, our findings with *D. discoideum* might apply universally to all slime molds, but the timing of when these steps in differentiation occur varies. Whatever the ultimate answer to this question, we have learned many interesting things about the process of the formation of presumptive stalk cells and spores. It is a step in development that has been amenable to observation and experimentation.

The New Anatomy

I am always stuck by how important new techniques in biology will open up totally new vistas of research. This is nowhere more evident in molecular biology and what we have learned using it. One of the advances that has played a big role in the study of slime mold differentiation (and in many other organisms as well) is to find a protein that is characteristic of one cell type, for instance stalk cells (or spores), and attach the gene for the green fluorescent protein (originally discovered in a jelly fish) to the promoter of the slime mold cell-specific gene. In this way one can then observe and follow the movements and the development of a specific cell type. As is obvious, this ability has many advantages over vital dyes as a method of studying what is going on in the zones, and has been elegantly exploited by Jeffrey Williams and his associates and others.[101]

The first thing they discovered that was of major importance is that there are more than just two zones, but subzones—giving a fuller picture of the differentiation process. The anterior zone can be partitioned into three parts that express different prestalk genes. As the fruiting body rises up into the air, the top and the bottom of the prespore region has zones called the upper and lower cup. This is indeed a triumph for using molecular biology in the pursuit of slime mold development.

Another remarkable discovery was based on the earlier work of John Sternfeld and Charles David, who found that some of the amoebae that were distributed throughout the prespore region stained darkly with vital dyes. They called them anterior-like cells.[102] When molecular labeling became possible, it was discovered that even though they are in the prespore region, they expressed prestalk genes and that they

---

[101] For a brief review, see Kessin (2001), pp. 143 ff.
[102] Sternfeld and David (1982).

move forward to become the upper and lower cups. Their origin was shown to be by the conversion of individual pre-spore cells into anterior-like cells.

CELL SORTING AND DIFFERENTIATION

This brings us to the role of cell movements involved in the process of differentiation, and the activities of anterior-like cells are a perfect example. We have already seen in the discussion of slug movement that it involves the contribution of the motion of all the amoebae, and furthermore we know that individual amoebae can move independently of one another. Remember the case of the vitally stained slugs that turned around in a dead-end agar tunnel where the anterior red cells percolated through the colorless posterior cells to reemerge at the opening of the tunnel (see fig. 22).

This raises a fascinating and fundamental question common to all of developmental biology. Is the fate of a cell the result of its position in the whole embryo (or slug), or is the fate of a cell something that is predetermined from its onset? The former is a regulative development, first made prominent by the famous experiments of Hans Driesch in the nineteenth century.[103] To put the matter in its simplest, detail-free form, he cut a sea urchin embryo in two, and both halves reorganized to produce two perfect, but dwarf larvae. The fate of a cell is a function of its position in the whole. The only regrettable part of this story is that it amazed Driesch to such a degree that he argued that any mechanical explanation was impossible and that a vital force must be doing this remarkable regulation. One is certainly reminded

[103] Driesch (1907).

of the present-day creationist arguments, and in particular that of intelligent design, where rational explanations are ruled impossible before even looking for them.

Roughly at the same time in the nineteenth century, E. B. Wilson and independently E. G. Conklin in the United States discovered that invertebrates from some other groups had a radically different kind of development. Each part of the early embryo, beginning after the fertilization of the egg, was mapped out from the very beginning. This meant that if a portion of the embryo that was destined to be a specific part of the larva was removed, the larva would be lacking that part. These two different kinds of animal development were called "regulative," for they rearranged themselves internally to come out complete and whole, and "mosaic" where fate and position is established right from the beginning. It has long been realized that both may occur in the same organism, and that they are not an all-or-none phenomenon.

If one looks at the cellular slime molds with these two seemingly antithetic ways of developing, one can clearly see that both are involved, but in a very interesting way. Let me explain.

It should be said right from the beginning that cellular slime molds have vast powers of regulation. As Kenneth Raper showed in his remarkable paper of 1940 (which really inspired all the subsequent work on the development of slime molds), if a slug is cut into pieces, each piece will produce a fruiting body (fig. 30).[104] In the anterior zone, which was all prestalk cells, once cut into a segment, some of the amoebae reverted

[104] Raper (1940b).

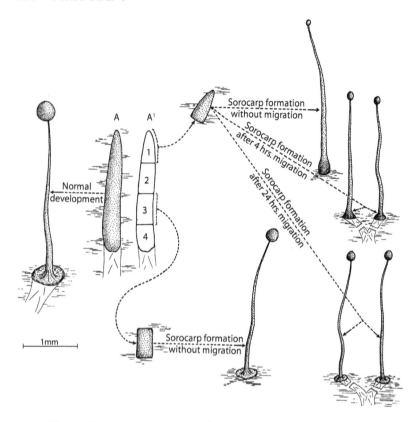

Fig. 30. Fruiting bodies from a large slug cut into segments. (From Raper 1940)

to prespore cells. He showed that the process required time, but that if the migration after cutting was sufficient, the spore-stalk ratio would be normal in the subsequent fruiting body; but if fruiting occurred soon after the operation, the resulting fruiting body had a disproportionately large stalk and small sorus. It takes time for a prestalk cell to transform into a prespore cell, but it can readily take place.

In a posterior segment made up of prespore cells, the conversion appears to take place rapidly, and now we know why. As mentioned earlier, a few prespore cells in the posterior are converted into prestalk cells: they are the anterior-like cells with their prestalk proteins, and the moment the segment loses its tip and a new dominant tip is formed, these amoebae rush forward and establish a new prestalk zone. The supply of anterior-like cells is eventually replenished in the posterior region by the conversion of additional prespore cells. So it is doubtful that the prespore-to-prestalk transformation is really more rapid than the prestalk-to-prespore one, for the process of the creation of new anterior-like cells might well also be slow.

There is another fact that fits in here neatly. If the Raper experiment of cutting a slug into segments is done on slugs that have migrated some distance, his result is upheld. However, if it is performed on a new slug that has just formed, then all the segments, including the anterior one, produce normally proportioned fruiting bodies without delay.[105] In other words, if the progression toward the advanced prestalk state has not yet occurred, there is no need for a slow reversing of the partial differentiation.

The important conclusion from these observations is that cellular slime molds are highly regulative, and that depending upon their position in the slug they can convert to the appropriate presumptive cell type. Any individual amoeba can shift the direction of its development in the stalk or spore direction—they are readily interchangeable—and what determines their fate is the position of the amoeba in the whole. The slug is, in the flowery language of Hans Driesch,

---

[105] Bonner and Slifkin (1949).

"a harmonious equipotential system." I am never quite sure what that means, and I would translate it into more modest words by saying that the slug is a unified whole that has different regions that determine the fate of the cells within. Such a system relies entirely on the fact that one can convert from one cell type to another, either forward or backward. The virtue of cellular slime molds for experimental analysis is that there are only two cell types and not many as in most other multicellular organisms.

How, after all I have just said, can we say that there is anything mosaic about the development of slime molds? This in itself makes a very interesting story. Right from the beginning I should point out that we are not talking about a rigid, permanently fixed mosaic development, for the ability of interconversion of the two cell types just discussed will not permit it. Instead we should think of amoebae at different stages of development as showing tendencies towards one of the two cell fates. They are predilections that are always capable of reversion.

The amoebae at the beginning of aggregation are not identical: they vary in size, in density, in the amount of food energy stored, and in other respects no doubt related to these. What happens as they aggregate and migrate is that some of these cells will move to the anterior end of the cell mass, and others will lag behind. They seem right from the beginning to be predisposed toward becoming a stalk cell or a spore—at least they are leaning in one of these directions. But as I have just said, they are not fixed in their fate, and at any time, depending on the circumstances, they can convert from one cell type to the other. The important point is that their position in the multicellular stage of their life cycle is not fixed;

as individual cells they can move about within the mass; there can be cell sorting.

The fact that amoebae could sort out independently was not obvious for some years despite an early experiment that seemed to suggest it. If vitally stained anterior cells were grafted onto the posterior of an unstained slug, during the course of the next few hours the stained cells slowly moved forward in a band (again by infiltrating through other amoebae) to eventually become part of the tip of the previously colorless slug (fig. 31). Clearly, the stained anterior amoebae returned to their original position; they sorted out, albeit in a large group. That individual cells did this was shown by the sorting out of mutant cells in a *D. discoideum* slug and in the whorls of *Polysphondylium*.[106] If the cells from a normal strain and those of an easily identifiable mutant that produces an aberrant fruiting body were mixed before aggregation, it could be shown that the mutant amoebae were not evenly spaced along the slugs (or the whorls), but that they were present in varying proportions along the axis of the slug. In other words, the mutant amoebae moved at a different rate from the wild-type amoebae.

The next big advance came when it was discovered by John Ashworth and his associates that one could treat amoebae so that they would be the ones to end up either fore or aft in the slug.[107] By raising the amoebae on either a rich supply of food, or feeding them on a minimal diet (and if the cells were marked appropriately so they could be followed), then the well-fed amoebae ended up in the posterior

---

[106] Bonner (1959a,b).
[107] Leach et al. (1973).

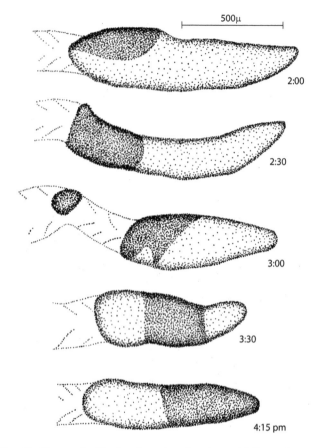

Fig. 31. Camera lucida drawings showing the rapid forward movement of a vitally stained anterior portion which has been grafted into the posterior region of an intact migrating slug. (From Bonner 1952)

prespore region, and the deprived amoebae became prestalk amoebae. This seemed to make sense, for spores could use the energy stores for the next generation, while prestalk cells only needed enough energy for building the stalk, and then they die.

The next interesting discovery in this saga was made by Anthony Durston and S. A. McDonald.[108] They showed that the fate of an amoeba depended on when, in its cell cycle, the amoeba began its preaggregation starvation. If it was deprived of food just before cell division, when the amoeba is replete, it becomes a prespore amoeba; if starvation happens just after cell division when it is leanest, then it becomes a prestalk amoeba. Both of these sets of experiments tell us the same thing: well-fed amoebae move in the spore direction, while the less well fed ones move in the stalk direction. This conclusion is supported by an earlier observation whose significance was not understood at the time: from the beginning of migration on, there are significantly more residual cell divisions (mitoses) among the prespore amoebae in a slug than in the prestalk ones.[109] This is also consistent with the idea that prespore amoebae have more energy stores.

Differences in the amoebae can be seen prior to aggregation. It has been shown by R. Baskar that even spores, before they germinate, will show differences in their ability to stain with vital dyes such as Neutral Red.[110] What is especially remarkable is that 20 percent of these spores show the dye strongly and go to the prestalk region, while the remaining unstained spores end up in the posterior prespore region.

Let us now consider *Polysphondylium* in the light of all that has been said: how can it be so different and not show the presumptive regions in the same way yet, as we have just seen

[108] A number of papers on this subject have appeared following the work of McDonald and Durston (1984). See Kessin (2001) for a comprehensive review.

[109] Bonner and Frascella (1952); Durston and Vork (1978).

[110] Baskar (1996).

from the mutant studies, still have sorting out? Our understanding of this anomaly took a big leap forward as a result of some ingenious experiments done by Edward Cox and his fellow workers.[111] After fusing two *Dictyostelium* promoters to the fluorescent protein and inserting them into the *P. pallidum* genome, they found that a prestalk gene was randomly distributed in the beginning. Later, in the slug stage, it was in the anterior end, while a prespore gene in a similar experiment was concentrated posteriorly. The result was identical to that found in *D. discoideum*: the presumptive stalk and spore cells sorted out to their appropriate regions. This is so, yet in *Polysphondylium* most of the amoebae are prespore, and the prestalk amoebae only appear at the tip just before stalk formation. The time course of their differentiation is different from *D. discoideum*; one is tempted to think of the differences between the two as a difference in timing, and that the basic steps are the same for both.

Proportions

Right in the beginning Kenneth Raper pointed out that regardless of the size of the slug—and the size could vary enormously—the proportions of spores and stalk cells of a fruiting body were roughly constant.[112] Since then there have been many careful studies to find out how constant and accurate this proportion is.[113] It has to do with a basic human urge: we love to measure things and derive great satisfaction

---

[111] Vocke and Cox (1992); Gregg and Cox (2000).

[112] Raper (1941).

[113] See Nanjundiah and Bhogle (1995) for a detailed set of measurement and a review of the work of others.

from doing so. The results have provided two bits of information: one is that larger fruiting bodies have a slightly higher percentage of spores than smaller ones, and the other is that the ratios of stalk to spore vary and are to some degree approximate.[114] It is the latter that has produced some contention, but rather than join the argument, I want to ask a more fundamental question. Is the degree of accuracy important? My answer would be: it is not. However precise or imprecise the proportions might be, the key point is that there is a division between spore and stalk, and we want to know how this comes into being, how it is determined. Knowing the degree of variation will probably not be helpful because that in itself will probably not illuminate the mechanism. My plan is to first discuss the external conditions that shift the proportions, and then go into what we know of the actual mechanism that directs some amoebae to become spores and others stalk cells.

FACTORS THAT AFFECT PROPORTIONS

There are external factors that can affect the proportions. One is temperature, and if amoebae of D. discoideum are raised at a relatively cool temperature (17°C) and at the beginning of fruiting are shifted to a significantly warmer environment (27°C), the percentage of spores in the resulting fruiting bodies will greatly increase (from roughly 85 percent to 98 percent).[115] The difficulty with this curious fact is that it is hard

---

[114] Rafols et al. (2001).

[115] Bonner and Slifkin (1949). P. A. Farnsworth (1975) showed that the effect of a shift in temperature worked only if the shift occurred at the onset of fruiting. Shifts at earlier stages of development produced normal proportions.

to understand its significance: we do not know what the temperature shift does, how it acts. It does not seem to have any obvious explanation.

In many ways the effect of oxygen is the most interesting. The first to discover it was John Sternfeld, who showed that with an increase in the percent oxygen in the atmosphere above cultures of developing slugs that are vitally stained with Neutral Red, *D. discoideum* produced slugs with disproportionately large red prestalk zones.[116]

This has also been examined in amoebae drawn up into a capillary tube. Immediately (within a minute) after the cells are in place, the end of the capillary exposed to the oxygen takes on a deeper red if all the amoebae are stained with Neutral Red. However, it takes an hour or more for these anterior cells to show the presence of prestalk specific genes when using the technique of attaching a fluorescent gene to a prestalk promoter. In other words, the anterior zone in the capillary is manufacturing prestalk proteins, and time-lapse recordings show that, as in normal slugs, its amoebae are in rapid motion compared to the relatively passive posterior amoebae. As one might expect, further development is not possible in the confinement of the capillaries. That this effect is produced by high oxygen has been verified in other ways. If the capillary is exposed to different concentrations of oxygen, the higher the percentage of oxygen, the larger the rapidly established anterior zone, paralleling Sternfeld's result.[117]

There is also some evidence that a decrease in oxygen will hasten the appearance of prestalk and prespore zones in migrating slugs stained with Neutral Red. As discussed earlier,

[116] Sternfeld (1988).
[117] Bonner et al. (1995).

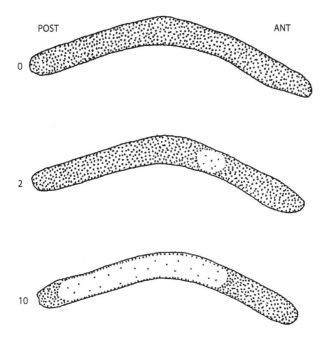

Fig. 32. Diagrammatic drawings based on a videotape showing the effect of submersion in mineral oil in hastening the beginning of differentiation. The prespore blanching begins after two minutes, and by ten minutes it has spread to its normal extent. (From Bonner et al. 1990)

these presumptive cells do not arise immediately at the end of aggregation, but may take up to five hours to do so. However, if an immature uniformly stained slug is submerged under mineral oil (which contains less oxygen than air, but more than water), then within ten minutes some of the prespore amoebae will appear in a small area at the edge of the future division line and then subsequently spread backwards over the entire prespore region (fig. 32). It is difficult to know how to interpret this, but again differences in oxygen concentration have a big effect on differentiation.[118]

The capillary experiments provide another key bit of information. In keeping with the experiments mentioned earlier, the amoebae are very sensitive to a gradient of oxygen. If a short capillary is the same diameter over its length, then the active prestalk zones will form at both ends and remain stable for a number of hours. However, if the capillaries are tapered ever so slightly, the anterior zone at the larger end will, in a matter of hours, slowly become dominant and take over (fig. 33). This minuscule difference in oxygen tension at the two ends is sufficient to establish the polarity of the trapped amoebae, that is, establish which end is the front and which is the rear.[119]

METABOLIC GRADIENTS

The oxygen experiment raises an ancient subject that has always fascinated me. When I was a student, a well-known developmental biologist, Charles Manning Child at the University of Chicago, spawned a theory of development that was simplicity itself.[120] He argued all developing organisms have a metabolic gradient where the high point consumed the most oxygen, and that consumption decreased along the axis of the embryo: there was a gradient in the rate of metabolism. He and his students demonstrated the reality of this gradient by experiments on many organisms. He considered this gradient of oxygen consumption to be the ultimate

[118] Bonner et al. (1990).
[119] Bonner et al. (1998).
[120] Child (1941).

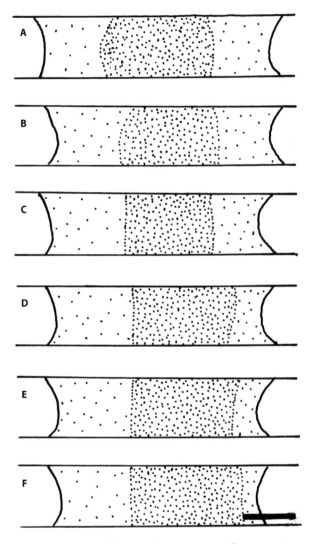

Fig. 33. Drawings traced from a videotape. A capillary containing slug amoebae and exposed to air at both ends. (From Bonner et al. 1998)
(A) An anterior zone at both ends after two minutes.
(B through F) Images gathered at sixty-minute intervals afterwards.
(Bar = 100 μm.)

driving force of all development. The interesting thing to me has always been the general dismissal of his ideas at the time and the ignoring of his views to some degree today.[121]

Earlier the main complaint was that his idea was too simple to be of any use in understanding the complex process of development. How could a gradient of metabolism, which is after all no more that a quantitative gradient, be responsible for all the diverse qualitative differences that arise by differentiation? Even as a youngster I always had a fondness for the underdog, especially when it involved a new and good idea. Today it seems particularly ironic that, starting with Alan Turing in the 1950s, we take for granted that through reaction-diffusion phenomena it is easy to see how simple gradients can lead to patterns and differences in parts during development. As a result there are many models of slime mold development, including some that give useful insights into how we could explain the various stages of development, and many of them involve reaction-diffusion processes.[122] But there is never any mention of Charles Manning Child or of metabolic gradients, yet they are the foundation of all of them.

Slime molds and classical experiments on metabolic gradients are connected in another interesting way. Many of the crucial experiments were done on hydroids, especially *Tubularia*, which consists of a cylindrical stem (made of chitin) with a terminal mouth and tentacles, the hydranth. If the cylinder is cut into segments, each segment will regenerate a new hydranth at its anterior end. T. H. Morgan, the father of modern genetics, showed in an early experiment that if

[121] See Neil Blackstone (2006) for a historical discussion of this phenomenon (and further references).

[122] For example, Meinhardt (1982).

these bits of stem were imbedded in sand at the bottom of the aquarium, the hydranths always appeared at their upper end, no matter what their original orientation had been. Later it was shown that this occurred because there was more oxygen at their free ends than the end buried in the sand. (The plot thickened when it was shown that there was also an inhibitor that accumulated in a confined space. Eventually it was made clear that both oxygen and the inhibitor could determine the polarity of stem segments.) The role of oxygen was used as evidence for the role of metabolic gradients in hydroid regeneration.[123]

What strikes me as so interesting is that in the glass capillary experiment with slime molds described above, if one end of the capillary is slightly larger than the other, that end becomes the dominant one. The cylinder in this case is made of glass rather than chitin, but the parallel in the result seems to me quite remarkable. In both cases polarity is established by an oxygen gradient.

What presumably happens when slime molds develop under normal circumstances is that a few amoebae at the beginning of aggregation become particularly active metabolically; they become the first to secrete cyclic AMP in any quantity, and the tip remains dominant and maintains the polarity. All the other events that follow produce differentiation with its division line and subsequently with the proportional separation of stalk cells and spores. None of these considerations involving oxygen and metabolism tell us anything about those processes; they only make it possible by establishing the basic body plan. I should add that in his big book in which he summarizes his lifework,

---

[123] For reviews of all this early work on regeneration in hydroids, see Barth (1940) and Tardent (1963).

Child has a brief section on the cellular slime molds where he speculated—but provided no evidence—that his principles surely apply.

Next comes the big question. How do we get from a gradient to proportions?

THE MECHANISM OF DETERMINING PROPORTIONS

Unfortunately we do not have a definitive answer to the question above. We have many tantalizing bits of information, but they have not jelled into one answer. It is generally agreed that there are chemical messengers that travel from one end of the slug to the other. Furthermore, Kei Inouye put the two cell types on different sides of a dialysis membrane that can only allow small molecules to pass, and it was clear that the signaling substance (or substances) that is produced by spores would pass through and prevent prestalk cells from turning into prespore amoebae.[124] Among other things this meant that the cells did not need to touch one another for the effect to succeed; the diffusion of a small molecule is quite sufficient. What was this key molecule, or is there more than one?

The matter took a big step forward when Robert Kay and his associates found a substance that seemed to fill the bill.[125] It was a small compound (containing a chlorine atom) that he called the Differentiation Inducing Factor, and fortunately

---

[124] Inouye (1998).

[125] The DIF literature is reviewed in Kessin (2001). It is now known that there is more than one DIF and the others differ in their structure from the original one. If the DIF mechanism is suppressed using molecular techniques, stalk cell differentiation will still occur. There is a jungle of facts that still needs to be sorted out.

he gave it the acronym DIF. It was a substance that specifically induced stalk cells and was produced by the prespore cells, as though the prespore cells were forcing the anterior cells to become stalk cells. As can be immediately seen, this has a very attractive logic to it. The stalk cells die—they are the true altruists—but now it appears that the prespore amoebae, by producing DIF that diffuses to the anterior, are forcing their neighbors to sacrifice themselves. (The very same phenomenon is found among the social insects, which show a balance in different castes. For instance, termites have a fixed ratio for the numbers of workers and soldiers, and it is known that the ratio is maintained by one caste that gives off a substance that controls the number—or percentage—of individuals of the other caste.[126] The whole idea is very appealing, for it makes sociobiological sense.) Unfortunately, in the case of slime molds the situation might be far more complicated, and it has not yet been sorted out.

The main difficulty is that there are a number of substances that might be involved in the determination of the presumptive stalk and spore zones—substances that are known to be produced by amoebae that inhibit the transformation of prestalk amoebae into prespore ones. One knows very little about their location in different parts of the slug, but they are possible candidates for doing something that is similar to what was just postulated for DIF. And to complicate matters, there is more than one DIF; there is a family of derivatives. They could be interacting with the DIFs in various complex ways, or they could be reinforcing the action of the DIFs. As I have said before, the idea that one can have different

---

[126] See Wilson's (1971) discussions of caste determination.

systems doing the same thing is not unusual. After all, if by mutation an important step in development is duplicated with a different chemical pathway, it will not be eliminated by natural selection. On the contrary, it is a safety measure to have an additional way of performing an important step that is producing a fruiting body supported by a stalk, something so necessary for spore dispersal.

8 THE FUTURE

In trying to give a general picture of what we know about slime molds, I have neglected much of the recent research. I have done this consciously because for almost every issue that I have brought up the plot keeps thickening. It has to do with the very nature of the molecular biology of slime molds today. For many of the phenomena I have discussed, and this is especially true of the discussion of their development, one wants to understand the molecular events that led to the final result. One wants to fully analyze all the events. For instance, take what has just been discussed: the differentiation of stalk cells and spores and how they come into being, how they are moved about (their morphogenesis) to produce their final form, and how their proportions are maintained. For each of these events one looks for the genes that are responsible, and this can be done by using various clever techniques. One can block the action of a gene and see if there is something missing in the development that required that gene. This has been especially successful in the case of genes for the binding of cyclic AMP discussed earlier,

and how this initial event is relayed to the internal cytoskeleton. In this way one can determine whether cyclic AMP is necessary for some developmental process. The tools of molecular biology are powerful, and we are beginning to understand many things about slime mold development at a more fundamental level. And we will continue to do so; this is clearly the way of the immediate future.

There is one aspect of this course that we are taking that gives me pause. We are delving further and further down the chain of causal events, and this means we are looking at finer and finer details at the risk of losing the big picture. That is one of my main motivations for writing this book: I want to remind everyone about *all* of slime mold biology.

Another problem also arises as we probe deeper. It turns out that there is a great plethora of genes and gene products (including signaling substances) that are all together. The more details one unearths, the more complex becomes our picture of any step in development. Part of the difficulty is that genes and their immediate products do not act alone, but have complex networks in which each component requires interaction with other components to produce a particular developmental event. It appears increasingly the case that we are rapidly moving away from the primeval one-gene, one-protein, one-simple-phenotype plot, but rather to systems biology that includes all these interesting cooperations and communications between genes and their products.

We are going to have to find new ways of dealing with these complexities. It is the way of the future, and new ground will inevitably be broken. After all, the great goal in all of science is to find simple answers to big questions. The search for simple fundamental truths is what leads us forward and gives us a sense that we are gaining a deeper understanding of the world around us.

I suspect that one thing that may happen in all of developmental—or life cycle—biology is the increasing role of mathematics. At the moment we keep burrowing and finding more and more key genes and follow their often tortuous pathways to discrete developmental events. In other words, we can expect more and more details and an accumulation of concrete facts. This is absolutely necessary and definitely should be cheered forward, but it is not enough. These are the bricks that compose the mansion, but a far more profound question is, what is the nature of the architect and how are the vast assembly of genes and the substances they spawn orchestrated. There must be some sort of master plan that has been carefully built up over millions of years, for otherwise all the facts we are accumulating would not produce a slime mold, but only chaos.

We must turn that jungle into the simplicity that gives us the feeling that we are making scientific progress. And clearly one way—perhaps the only way—we can achieve this is through mathematical insight. The example we might look to for the future to understand the causal mechanisms in the development and evolution of cellular slime molds might be found in ecology.

There too, in earlier days, was an overabundance of facts that needed to be put into some sort of order. Demystifying that (literal) jungle has been one of the great successes of modern biology. This was in large part achieved through the genius of Robert MacArthur, who, with simple mathematics, was able to bring to light underlying principles that unify all the overwhelming details. To give one example among his many successes he, with E. O. Wilson, illuminated the principles of island biogeography in a big flash by showing that a finite set of key factors—such as island size, distance from the mainland, rates of immigration and emigration, and so forth—were sufficient to account for what is actually found on islands.[127]

Indeed, these principles have no obvious relevance to slime mold development, where a different set of multiple facts needs sorting out. Nevertheless, we can see the beginning of an era of enlightenment for slime molds. As I have repeatedly pointed out, there is an initial foundation of simple processes that arose by natural selection, and over the great span of geological time the same selection forces reinforced that adaptation with a plethora of checks and balances. This thickening of the plot was in itself adaptive.

One way to approach the matter is to speculate what the ancestral adaptations might have been like before the tangled overlay of secondary adaptations arose; however hypothetical, it would give some clue as to what might be the fundamental basis of the developmental steps.[128] To give an example, consider how the separation of spores and stalk cells could have arisen in a primordial cellular slime mold. One possibility might be the setting up of the simple reaction-diffusion mechanism of Alan Turing.[129] The difficulty is that we no longer have access to that ancient creature, so we can only surmise. But it is an intriguing guess, and the day may come (if it is not already here) where we may hail Turing, along with his other claims to fame, as the MacArthur of developmental biology.

[127] MacArthur and Wilson (1967).
[128] I develop this theme further in Bonner (2000).
[129] Turing (1952).

Another approach is already emerging from the study of genomics and development. It is increasingly obvious that knowing all the genes identified in an organism does not really tell us how the organism is constructed or how it develops. Again, it is a pile of bricks and the architect's plans are hidden from us. In the first place, as I said earlier, these genes and their products interact with one another in ways we are only beginning to perceive. Not only that, but they do different things at different times, so the whole process of life cycle development has a key time element that is central. To state these obvious facts is easy, but how one deciphers, and organizes, and simplifies this overwhelming heap is a great difficulty. Perhaps the way out will be to make abstract mathematical arguments that simplify and give some order to them.

A distinction should be made between the far future and the immediate future. We may not be ready for those grand mathematical simplifications yet, and we must keep digging just as we are at the moment. There are so many new things that are awaiting discovery in the development of cellular slime molds. One of the big advances has been the mapping of the complete genome.[130] It allows us to discover what genes they share with other organisms, often distantly related, whose function has already been elucidated, and this information will help us understand their role in slime molds. We can still expect to discover new, key signal molecules and the genes that control them. There is plenty of room for a lot more fruitful digging.

As is obvious from everything I have said in this book, there is so much more to be learned about every aspect of

---

[130] Eichinger et al. (2005).

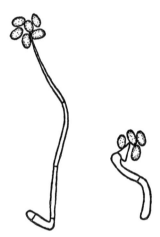

Fig. 34. The smallest known fruiting bodies. They are *Polysphondylium pallidum*. (From Bonner and Dodd 1962b)

the cellular slime molds. There is still much mystery to be found alongside the known. We are still miles away from getting a complete handle on what makes slime molds tick. I would like to end with one enigma that epitomizes for me what we do not yet know.

In the discussion of the stalk-spore proportions I wrote grandly about morphogens diffusing and somehow, perhaps by reaction-diffusion activities, determining how many stalk cells will be in the final fruiting body, and how many spores. This seems like a hopeful beginning to understanding the mechanisms involved. However, some years ago I was trying to produce very small fruiting bodies by supplying them with a minimum of bacterial food, and I managed to achieve the world's record. One fruiting body of *P. pallidum* consisted of three stalk cells and four spores, while the runner-up had five stalk cells and six spores (fig. 34). I have been assured by a physicist friend that these minute dimensions would make any diffusion-based system of producing proportions impossible, yet there they are! For me this gives a very strong message: we still have a long—and interesting—way to go.

And the reason we all started working on the cellular slime molds is that they were supposed to be so simple.

BIBLIOGRAPHY

Alvarez-Curto, E., D. E. Rozen, A. V. Ritche, C. Fouquet, S. L. Baldauf, and P. Schaap (2005) Evolutionary origin of cAMP-based chemoattraction in the social amoebae. *Proc. Nat. Acad. Sci. USA.* 102: 6385–6390.

Amagai, A. (1984) Induction by ethylene of macrocyst formation in the cellular slime mould *Dictyostelium mucoroides. J. Gen. Microbiol.* 130: 2961–2965.

Baker, J. R. (1948) The status of the protozoa. *Nature* 161: 548–551, 587–589.

Barth, L. G. (1940) The process of regeneration in hydroids. *Biol. Rev.* 15: 405–420.

Baskar, R. (1996) Early commitment of cell types and a morphogenetic role for calcium in *Dictyostelium discoideum*. Ph. D. *thesis*, Indian Institute of Science, Bangalore, India.

Blackstone, N. W. (2006) Charles Manning Child (1869–1954): The past, present, and future of metabolic signaling. *J. Exp. Zool. (Mol. Dev. Evo.)* 306B: 1–7.

Bonner, J. T. (1947) Evidence for the formation of cell aggregates by chemotaxis in the development of the slime mold *Dictyostelium discoideum. J. Exp. Zool.* 106: 1–26.

Bonner, J. T. (1949) The demonstration of acrasin in the later stages of the development of the slime mold *Dictyostelium discoideum. J. Exp. Zool.* 110: 259–272.

Bonner, J. T. (1952) The pattern of differentiation in the amoeboid slime molds. *Am. Nat.* 86: 79–89.

Bonner, J. T. (1959a) Evidence for the sorting out of cells in the development of the cellular slime molds. *Proc. Nat. Acad. Sci. USA* 45: 379–384.

Bonner, J. T. (1959b) Differentiation in social amoebae. *Scientific American* 201: 152–162.

Bonner, J. T. (1993) Proteolysis and orientation in *Dictyostelium* slugs. *J. Gen. Microbiol.* 139: 2319–2322.

Bonner, J. T. (1995) Why does slug length correlate with speed during migration in *Dictyostelium discoideum? J. Biosciences* 20: 1–6.

Bonner, J. T. (1998) A way of following individual cells in the migrating slugs of *Dictyostelium discoideum. Proc. Nat. Acad. Sci. USA.* 95: 9335–9395.

Bonner, J. T. (2000) *First Signals.* Princeton: Princeton University Press.

Bonner, J. T. (2001) A note on the number of cells in a slug of *Dictyostelium discoideum.* http://dictybase.org/bonner%20paper.pfd

Bonner, J. T. (2006) Migration in *Dictyostelium polycephalum. Mycologia* 98: 260–264.

Bonner, J. T., and M. S. Adams (1958) Cell mixtures of different species and strains of cellular slime moulds. *J. Embryol. Exp. Morphol.* 6: 346–356.

Bonner, J. T., and M. R. Dodd (1962a) Evidence for gas-induced orientation in the cellular slime molds. *Develop. Biol.* 5: 344–361.

Bonner, J. T., and M. R. Dodd (1962b) Aggregation territories in the cellular slime molds. *Biol. Bull.* 122: 13–24.

Bonner, J. T., and D. Eldredge, Jr. (1945) A note on the rate of morphogenetic movement in the slime mold *Dictyostelium discoideum. Growth* 9: 287–297.

Bonner, J. T., and E. B. Frascella (1952) Mitotic activity in relation to differentiation in the slime mold *Dictyostelium discoideum. J. Exp. Zool.* 121: 561–572.

Bonner, J. T., and D. S. Lamont (2005) Behavior of cellular slime molds in the soil. *Mycologia* 97: 178–184.

Bonner, J. T., and M. K. Slifkin (1949) A study of the control of differentiation: The proportions of stalk and spore cells in the slime mold *Dictyostelium discoideum. Amer. J. Bot.* 36: 727–734.

Bonner, J. T., W. W. Clarke, Jr., C. L. Neely, Jr., and M. K. Slifkin (1950) The orientation to light and the extremely sensitive orientation to temperature gradients in the slime mold *Dictyostelium discoideum. J. Cell. Comp. Physiol.* 36: 149–158.

Bonner, J. T., B. D. Joyner, A. Moore, H. B. Suthers, and J. A. Swanson (1985) Successive asexual life cycles of starved amoebae in the cellular slime mold, *Dictyostelium mucoroides* var. *stoloniferum. J. Cell. Sci.* 76: 23–30.

Bonner, J. T., H. B. Suthers, and G. M. Odell (1986) Ammonia orients cell masses and speeds up aggregating cells of slime moulds. *Nature* 323: 630–632.

Bonner, J. T., A. Chiang, L. Lee and H. B. Suthers (1988) The possible role of ammonia in phototaxis of migrating slugs of *Dictyostelium discoideum. Proc. Nat. Acad. Sci. USA* 85: 3885–3887.

Bonner, J. T., I. N. Feit, A. K. Selasse, and H. B. Suthers (1990) Timing of the formation of the prestalk and prespore zones in *Dictyostelium discoideum. Develop. Genetics.* 11: 439–441.

Bonner, J. T., K. B. Compton, E. C. Cox, P. Fey, and K. Y. Gregg (1995) Development in one dimension: The rapid differentiation of *Dictyostelium discoideum* in glass capillaries. *Proc. Nat. Acad. Sci. USA* 92: 8249–8253.

Bonner, J. T., L. Segel, and E. C. Cox (1998) Oxygen and differentiation in *Dictyostelium discoideum. J. Biosci.* 23: 177–184.

Bonner, J. T., P. Fey, and E. C. Cox (1999) Expression of prestalk and prespore proteins in minute, two-dimensional *Dictyostelium* slugs. *Mechanisms of Development* 88: 253–254.

Buder, J. (1920) Neue phototropische Fundamentalversuche. *Ber. Deutsch Bot. Gaz.* 38: 10–19.

Buss, L. W. (1982) Somatic cell parasitism and the evolution of somatic tissue compatibility. *Proc. Nat. Acad. Sci. USA* 79: 5337–5341.

Cavender, J. C. (1973) Geographical distribution of Acrasieae. *Mycologia* 65: 1044–1054.

Cavender, J. C., and S. L. Stephenson (2002) Dictyostelid cellular slime moulds in the forests of New Zealand. *New Zealand J. Botany* 40: 235–264.

Child, C. M. (1941) *Patterns and Problems of Development.* Chicago: University of Chicago Press.

Cox, E. C., F. Spigel, G. Byrne, J. W. McNally, and L. Eisenbud (1988) Spatial patterns in the fruiting bodies of the cellular slime mold *Polysphondylium pallidum. Differentiation* 38: 73–81.

Dao, D. N., R. H. Kessin, and H. Ennis (2000) Developmental cheating and the evolutionary biology of *Dictyostelium* and *Myxococcus. Microbiology* 146: 1505–1512.

Dorman, D., and C. J. Weijer (2001) Propagating chemoattractant waves coordinate periodic cell movement in *Dictyostelium* slugs. *Development* 128: 4535–4543.

Driesch, H. (1907) *The Science and Philosophy of the Organism.* London: Black.

Durston, A. J. (1976) Tip formation is regulated by an inhibitor gradient in the *Dictyostelium discoideum* slug. *Nature* 263: 126–129.

Durston, A. J., and F. Vork (1978) The spatial pattern of DNA synthesis in *Dictyostelium discoideum* slugs. *Exp. Cell Res.* 115: 454–457.

Eichinger, L., et (numerous) al.'s (2005) Genome of the social amoeba *Dictyostelium discoideum. Nature* 435: 43–57.

Ennis, H. L., D. N. Dao, S. U. Pukatzki, and R. H. Kessin (2000) *Dictyostelium* amoebae lacking an F-box protein form spores rather than stalk in chimeras with wild type. *Proc. Nat. Acad. Sci. USA* 97: 3292–3297.

Fankhauser, G. (1955) The role of the nucleus and cytoplasm. In *Analysis of Development*, pp. 126–150. B. H. Willier, P. A Weiss, and V. Hamburger, eds. Philadelphia: W. B. Saunders.

Farnsworth, P. A. (1975) Proportionality in the pattern of differentiation of the cellular slime mould *Dictyostelium discoideum. J. Embryol. Exp. Morph.* 33: 869–877.

Filosa, M. F. (1962) Heterocytosis in cellular slime molds. *Am. Nat.* 96: 79–91.

Fisher, P. (1991) The role of gaseous metabolites in phototaxis by *Dictyostelium discoideum* slugs. *FEMS Microbiol. Lett.* 77: 117–120.

Fisher, P. R., W. N. Dohrmann, and K. L. Williams (1984) Signal processing in *Dictyostelium discoideum* slugs. In *Modern Cell Biology,* B. H. Satir, ed., 197–248. New York: A. R. Liss.

Francis, D. W. (1964) Some studies on phototaxis of *Dictyostelium. J. Cell Compar. Physiol.* 64: 131–138.

Gerish, G., D. Malchow, V. Riedel, E. Müller, and M. Every (1972) Cyclic AMP phosphodiesterase and its inhibitor in slime mold development. *Nature New Biology* 235: 90–92.

Gomer, R. H. (1999) Cell density sensing in a eukaryote. *ASM News* 65: 23–29.

Gregg, K., and E. C. Cox (2000) Spatial and temporal expression of a *Polysphondylium* spore specific gene. *Develop. Biol.* 224: 81–95.

Häder, D.-P., and A. Hansel (1991) Responses of *Dictyostelium discoideum* to multiple environmental stimuli. *Bot. Acta* 104: 200–205.

Hagiwara, H. (1989) The taxonomic study of Japanese Dictyostelid cellular slime molds. Tokyo: National Science Museum Press.

Haldane, J.B.S. (1955) Some alternatives to sex. *New Biology* 19: 7–26.

Hohl, H. R., and K. B. Raper (1964) Control of sorocarp size in the cellular slime mold *Dictyostelium discoideum. Dev. Biol.* 9: 137–153.

Hubbell, S. (2001) *The Unified Neutral Theory of Biodiversity and Biogeography.* Princeton, N.J: Princeton University Press.

Huss, M.J. (1989) Dispersal of cellular slime molds by two soil invertebrates. *Mycologia* 81: 677–682.

Inouye, K. (1998) Control of cell type proportions by a secreted factor in *Dictyostelium discoideum. Development* 107: 605–610.

Inouye, K., and I. Takeuchi (1979) Analytical studies on migrating movement of the pseudoplasmodium of *Dictyostelium discoideum*. *Protoplasma* 99: 289–304.

Inouye, K., and I. Takeuchi (1980) Motive force of the migrating pseudoplasmodium of the cellular slime mold *Dictyostelium discoideum*. *J. Cell Sci.* 41: 53–64.

Kahn, A. J. (1964) Some aspects of cell interaction in the development of the slime mold *Dictyostelium purpureum*. *Develop. Biol.* 9: 1–19.

Kaushik, S., and V. Nanjundiah (2003) Evolutionary questions raised by cellular slime mold development. *Proc. Indian Natl. Sci. Acad.* B69: 825–852.

Kaushik, S., B. Katoch, and V. Nanjundiah (2006) Social behaviour in genetically heterogeneous groups of *Dictyostelium giganteum*. *Behav. Ecol. and Sociobiol.* 59: 521–530.

Keating, M. T., and J. T. Bonner (1977) Negative chemotaxis in cellular slime molds. *J. Bact.* 130: 144–147.

Keller, E. F. (1983) The force of the pacemaker concept in theories of aggregation in cellular slime mold. *Perspectives in Biol. and Med.* 26: 515–521.

Kessin, R. H. (2001) *Dictyostelium*. Cambridge: Cambridge University Press.

Kessin, R. H., G. G. Gundersen, V. Zaydfudim, M. Grimson, and R. L. Blanton (1996) How cellular slime molds evade nematodes. *Proc. Nat. Acad. Sci. USA* 93: 4857–4861.

Konijn, T. M., K. van de Meene, J. T. Bonner, and D. S. Barkley (1967) The acrasin activity of adenosine-3', 5'-cyclic phosphate. *Proc. Nat. Acad. Sci. USA* 58: 1152–1154.

Konijn, T. M., D. S. Barkley, Y. Y. Chang, and J. T. Bonner (1968) Cyclic AMP: A naturally occurring acrasin in the cellular slime molds. *Am. Nat.* 102: 225–233.

Kopachik, W. (1982) Size regulation in *Dictyostelium*. *J. Emb. Exp. Morph.* 68: 23–25.

Kosugi, T., and K. Inouye (1989) Negative chemotaxis to ammonia and other weak bases by migrating slugs of the cellular slime molds. *J. Gen. Microbiol.* 135: 1589–1598.

Leach, C. K., J. M. Ashworth, and D. R. Garrod (1973) Cell sorting out during the differentiation of mixtures of metabolically distinct populations of *Dictyostelium discoideum*. *J. Embryol. Exp. Morphol.* 29: 647–661.

MacArthur, R. H., and E. O. Wilson (1967) *Island Biogeography.* Princeton, N.J.: Princeton University Press.

Macko, M. (1971) A comparison of diploid and haploid amoebae and pseudoplasmodia of the cellular slime mold *Dictyostelium discoideum* with respect to size and rates of movement. Senior thesis, Princeton University.

McDonald, S. A., and A. J. Durston (1984) The cell cycle and sorting behaviour in *Dictyostelium discoideum*. *J. Cell Sci.* 66: 195–204.

McMahon, T. A. (1971) Rowing: A similarity analysis. *Science* 173: 349–351.

McNally, J. G., J. G. Byrne, and E. C. Cox (1987) Branching in Polysphondylium whorls: Two-dimensional patterning in a three-dimensional system. *Dev. Biol.* 119: 302–304.

Meinhardt, H. (1982) *Models of Biological Pattern Formation.* New York: Academic Press.

Miura, K., and F. Siegert (1988) Light affects cAMP signaling and cell movement activity in *Dictyostelium discoideum. Proc. Nat. Acad. Sci. USA* 97: 2111–2116.

Mizutani, A., and K. Yanisagawa (1990) Cell-division inhibitor produced by the killer strain of cellular slime mold *Polysphondylium pallidum. Dev. Growth and Differ.* 32: 397–402.

Mizutani, A., H. Hagiwara, and K. Yanagisawa (1990) A Killer factor produced by the cellular slime mold *Polysphondylium pallidum. Arch. Microbiol.* 153: 413–416.

Nanjundiah, V., and A. S. Bhogle (1995) The precision of regulation on *Dictyostelium discoideum*: Implications for cell type proportioning in the absence of spatial pattern. *Indian J. Biochem. and Biophys.* 32: 404–416.

Olive, L. S. (1975) *The Mycetozoans.* New York: Academic Press.

Odell, G., and J. T. Bonner (1986) How the *Dictyostelium discoideum* grex crawls. *Phil. Trans. Roy. Soc. London* B312: 487–525.

Parent, C. A., and P. N. Devreotes (1996) Molecular genetics of signal transduction in *Dictyostelium. Ann. Rev. Biochem.* 65: 411–440.

Poff, K. L., and M. Skokut (1977) Thermotaxis by pseudoplasmodia of *Dictyostelium discoideum. Proc. Nat. Acad. Sci. USA* 74: 2007–2010.

Rafols, I. A., Y. Amagai, H. K., Maeda, H. K. MacWilliams, and Y. Sawada (2001) Cell type proportioning in *Dictyostelium* slugs: Lack of regulation within a 2.5-fold tolerance range. *Differentiation* 67: 107–116.

Raper, K. B. (1940a). The communal nature of the fruiting process in the *Acrasieae. Am. J. Bot.* 27: 436–448.

Raper, K. B. (1940b) Pseudoplasmodium formation and organization in *Dictyostelium discoideum. J. Elisha Mitchell Sci. Soc.* 56: 241–282.

Raper, K. B. (1941) Developmental patterns in simple slime molds. *Growth* 5: 41–76.

Raper, K. B. (1956) *Dictyostelium polycephalum* n. sp.: A new cellular slime mould with coremiform fructifications. *J. Gen. Microbiol.* 14: 716–732.

Raper, K. B. (1984) *The Dictyostelids.* Princeton, N.J.: Princeton University Press.

Raper, K. B., and M. S. Quinlan (1958) *Acytostelium leptosomum*: A unique cellular slime mould with an acellular stalk. *J. Gen. Microbiol.* 18: 16–32.

Raper, K. B., and C. Thom (1941) Interspecific mixtures in the Dictyosteliaceae. *Am. J. Botany* 28: 69–78.

Runyon, E. H. (1942) Aggregation of separate cells of *Dictyostelium* to form a multicellular body. *Collecting Net.* 17: 88.

Samuel, E. W. (1961) Orientation and rate of locomotion of individual amebas in the life cycle of the cellular slime mold *Dictyostelium mucoroides*. *Develop. Biol.* 3: 317–335.

Santorelli, L.A., C.R.L Thompson, E. Villegas, J. Svetz, C. Dinh, A. Parikh, R. Sucgang, A. Kuspa, J. E. Strassmann, D. C. Queller, and G. Shaulsky (2008) Facultative cheater mutants reveal the genetic complexity of cooperation in social amoebae. *Nature* 451: 1107–1110.

Satish, N., S. Sultana, and V. Nanjundiah (2007) Diversity of soil fungi in tropical deciduous forest in Mudumalai, southern India. *Current Science* 93: 669–677.

Schaap, P., T. Winkler, M. Nelson, E. Alvarez-Curto, B. Elgie, H. Hagiwara, J. Cavender, A. Milano-Curto, D. E. Rozen, T. Dingermann, R. Mutzel, and S. L. Baldauf (2006) Molecular phylogeny and evolution of morphology in the social amoebas. *Science* 314: 661–663.

Schindler, J., and M. Sussman (1977) Ammonia determines the choice of morphogenesis pathways in *Dictyostelium discoideum*. *J. Mol. Biol.* 116: 161–169.

Shaffer, B. M. (1956) Acrasin, the chemotactic agent in cellular slime moulds. *J. Exptl. Biol.* 33: 645–657.

Shaffer, B. M. (1957) Aspects of aggregation in cellular slime molds. *Am. Nat.* 91: 19–34.

Shaffer, B. M. (1961) The cells founding aggregation centres in the slime mould *Polysphondylium violaceum*. *J. Exptl. Biol.* 38: 833–849.

Shimomura, O., H.L.B. Suthers, and J. T. Bonner (1982) The chemical identity of the acrasin of the cellular slime mold *Polysphondylium violaceum*. *Proc. Nat. Acad. Sci. USA* 79: 7376–7379.

Siegert, F., and C. J. Weijer (1992) Three-dimensional scroll waves organize *Dictyostelium* slugs. *Proc. Nat. Acad. Sci. USA* 89: 6433–6437.

Spiegel, F. W., and E. C. Cox (1980) A one dimensional pattern in the cellular slime mould *Polysphondylium pallidum*. *Nature* 286: 806–807.

Stephenson, S. L., and J. C. Landholt (1992) Vertebrates as vectors of cellular slime moulds in temperate forests. *Mycol. Res.* 96: 670–672.

Sternfeld, J. (1988) Proportion regulation in *Dictyostelium* is altered by oxygen. *Differentiation* 37: 173–179.

Sternfeld, J., and C. N. David (1981) Oxygen gradients cause pattern orientation in *Dictyostelium* cell clumps. *J. Cell Sci.* 50: 9–17.

Sternfeld, J., and C. N. David (1982) Fate and regulation of anterior-like cells in *Dictyostelium* slugs. *Develop. Biol.* 93: 111–118.

Strassmann, J. E., and D. C. Queller (2007) Altruism among amoebas. *Natural History,* September, 24–26.

Suthers, H. B. (1985) Ground-feeding migratory songbirds as cellular slime mold distribution vectors *Oecologia* 65: 526–530.

Swanson, A. R., E. M. Vadell, and J. C. Cavender (1999) Global distribution of forest soil dictyostelids. *J. Biogeography* 26: 133–148.

Tardent, P. (1963) Regeneration in the Hydrozoa. *Biol. Rev.* 38: 293–333.

Thadani, V., P. Pan, and J. T. Bonner (1977) Complementary effects of ammonia and cyclic AMP on aggregation territory size in the cellular slime mold *Dictyostelium mucoroides*. *Exper. Cell. Res.* 108: 75–78.

Turing, A. M. (1952) The chemical basis of morphogenesis. *Phil. Trans. Roy. Soc. London* B237: 37–72.

Vocke, C. D., and E. C. Cox (1992) Establishment and maintenance of stable spatial patterns in lacZ fusion transformants of *Polysphondylium pallidum*. *Development* 115: 59–65.

Waddell, D. R. (1982) A predatory slime mold. *Nature* 298: 464–466.

Whitaker, B. D., and K. L. Poff (1980) Thermal adaptation of thermosensing and negative thermotaxis in *Dictyostelium*. *Exp. Cell Res.* 128: 87–93.

Wilson, E. O. (1971) *The Insect Societies*. Cambridge. Mass.: Harvard University Press.

Yamamoto, M. (1977) Some aspects of behavior of the migrating slug of the cellular slime mold *Dictyostelium discoideum*. *Develop. Growth and Differ.* 17: 93–102.

Yumura, S., K. Kurata, and T. Kitanishi-Yumura (1992) Concerted movement of prestalk cells in migrating slugs of *Dictyostelium* revealed by the localization of myosin. *Develop. Growth and Differ.* 34: 319–328.

Index